T0258738

SCIENCE
and the
STRUCTURE
of ETHICS

Transaction Books by Abraham Edel

Analyzing Concepts in Social Science

Anthropology and Ethics (with May Edel)

Aristotle and His Philosophy

Ethical Judgment

Exploring Fact and Value

In Search of the Ethical

Interpreting Education

Method in Ethical Theory

Relating Humanities and Social Thought

Science and the Structure of Ethics

SCIENCE
and the
STRUCTURE
of ETHICS

Abraham Edel

with a new introduction by
Irving Louis Horowitz

Routledge
Taylor & Francis Group

LONDON AND NEW YORK

Originally published in 1961 by The University of Chicago Press.

Published 1998 by Transaction Publishers

Published 2017 by Routledge
2 Park Square, Milton Park, Abingdon, Oxon OX14 4RN
711 Third Avenue, New York, NY 10017, USA

First issued in paperback 2018

Routledge is an imprint of the Taylor & Francis Group, an informa business

Copyright © 1998 by Taylor & Francis.

All rights reserved. No part of this book may be reprinted or reproduced or
utilised in any form or by any electronic, mechanical, or other means, now
known or hereafter invented, including photocopying and recording, or in any
information storage or retrieval system, without permission in writing from the
publishers.

Notice:
Product or corporate names may be trademarks or registered trademarks, and are
used only for identification and explanation without intent to infringe.

Library of Congress Catalog Number: 97-27083

Library of Congress Cataloging-in-Publication Data

Edel, Abraham, 1908–
   Science and the structure of ethics / Abraham Edel ; with a new in-
troduction by Irving Louis Horowitz.
     p. cm.
   Originally published: Chicago : University of Chicago Press, 1961, in
series: International encyclopedia of unified science.
   Includes bibliographical references.
   ISBN 1-56000-348-0 (cloth : alk. paper)
   1. Ethics. 2. Science—Moral and ethical aspects. 3. Science—
Philosophy. I. Title.
BJ57.E34   1997
170'.1'5—dc21   97-27083
                                                           CIP

ISBN 13: 978-1-138-51437-9 (pbk)
ISBN 13: 978-1-56000-348-9 (hbk)

Contents

ACKNOWLEDGMENTS   vii

INTRODUCTION TO THE TRANSACTION EDITION   ix

I. THE NATURE AND COMPLEXITY OF THE PROBLEM   1

     1. Issues in the "Relation of Science and Ethics"   1
     2. The Place of Scientific Results in Ethical Theory   2
     3. The Role of Scientific Method in Ethical Theory   4
     4. The Impact of the Scientific Temper in Ethical
         Theory   6
     5. Moralities   7
     6. Methodological Approaches in Ethics   8
     7. The Structure of an Ethical Theory   10

II. THE THEORY OF EXISTENTIAL PERSPECTIVES   11

     8. The Concept of an Existential Perspective (EP)   11
     9. Role of Existential Perspective within an Ethical
         Theory   15
     10. Overtly Scientific Existential Perspectives:
         Physical and Biological   17
     11. Overtly Scientific Existential Perspectives:
         Psychological   19
     12. Overtly Scientific Existential Perspectives:
         Sociocultural and Sociohistorical   23
     13. Science in Theological and Metaphysical
         Existential Perspectives   25
     14. Science in Transcendence Existential Perspectives   30
     15. Evaluation of Existential Perspectives   33

III. THE ROLE OF SCIENCE IN CONCEPTUAL AND METHODOLOGICAL
     ANALYSIS   45

     16. Conceptual-Methodological Frameworks   45
     17. Are There Workable Concepts of Moral
         Phenomena and Moral Experience?   48

18. Ethical Concept-Families and Their Existential
    Linkage   50
19. Organization, Generalization, Systematization   56
20. Validation, Verification, Reasoning, Justification   66
21. Application and Evaluative Processes   71

IV. DECISION, FREEDOM, AND RESPONSIBILITY   78

22. Evaluative Processes in Unstructured Situations   78
23. Toward a Scientific Study of Freedom and
    Responsibility   85
24. Toward a Strategy for Solution of the Free-Will
    Problem in Ethics   88
25. The Creative Temper and the Scientific Temper   92

NOTES   94

Acknowledgments

The present work originally appeared in 1961 as volume two, number three of the *International Encyclopedia of Unified Science,* published by the University of Chicago Press. Its republication is made possible through the kind action of Morris Philipson, director emeritus of the press, who turned over the copyright to me.

I should like also to acknowledge the work of Professor Irving Louis Horowitz in the introduction that follows. His intimate acquaintance with the vast literature of the social sciences over the intervening forty years has made it possible to appraise the work in the contemporary context. His introduction carries out this task and happily finds a place for its republication.

ABRAHAM EDEL
July 1997

Introduction to the Transaction Edition

Irving Louis Horowitz

> *Truth is the secret of eloquence and of virtue,*
> *the basis of moral authority;*
> *it is the highest summit of art and of life.*
> —Henri-Fréderic Amiel
> *(Journal Intime, 1883)*

Re-introducing a work first issued in 1961, and now offering the same effort to an audience in 1998 takes some explaining. While it may be the case that the course of philosophical growth, especially moral theory, runs slowly, the context in which any given position emerges, expands, and operates changes swiftly. As a consequence, the evaluation of a serious piece of work or a broad perspective changes, not so much as in science itself or as a consequence of discoveries that displace older paradigms, but as a result of new circumstances that make one question older visions.

Lest the reader think that Abraham Edel's volume *Science and the Structure of Ethics* falls into the category of a pleasant fossil, or that this starting point which emphasizes contexts rather than contents is a prelude to dismissal, let me at once reassure the reader that nothing could be further from the case, or from my mind. Indeed, it will be the burden of my commentary to show that Edel's volume is valuable precisely because it is remarkably resistant to fads and fashions of the moment—whether that moment be 1961 or 1998. The reason for this resiliency, in my opinion, is its *philosophical* centrism. Edel's monograph, and his work in general, is based on the tantalizingly simply proposition that moral knowledge should be neither mysterious nor pietistic. From the naturalistic tradition of Aristotle to the pragmatic tradition of Dewey, shared experience allows people to find out things for themselves.

But before we explore the contents of this prescient monograph, it is of some professional interest at least to know how this work came to be written. That Edel, clearly an outsider, and even a critic, of hard positivist circles, came to be asked to prepare this monograph for the prestigious, but often rigidly defined, *International Encyclopedia of Unified Science*, is itself part of the folklore of contemporary western philosophy. Happily, in this regard, and in response to an earlier conversation, the author has supplied the answer. It is an academic story well worth telling. Edel notes that

> Carnap, Neurath, and the rest of the positivists had their heart really in logic and epistemology. But to get a broader representation on the editorial board, they invited John Dewey. He wouldn't join them until they assured him that they did not share [A. J.] Ayer's view that value judgments were logically meaningless and really only emotive. So they added a book on values. They really had philosophy of art in mind. But when Charles Morris asked me to do that, I said that was my province of ignorance. He then said I should do it in ethics.

Whatever the idiosyncratic history of how this work came into existence, its extraordinary qualities as a synthesis of a naturalistic framework for describing and interpreting ethical frameworks cannot be denied.

Background factors notwithstanding, like so much of Edel's work, *Science and the Structure of Ethics* is rooted in historical context and human content. Edel understands well the relative nature of judgments, no matter how absolute and dedicated to ends a particular ethical position is held. What then was the context at the close of the remarkable decade of the 1950s—both in the broad cultural and in the narrower band identified as professional philosophy?

Despite the Holocaust and to a lesser extent the use of nuclear weapons in World War II, the West emerged from the period with its optimism intact. To be sure, one would have to admit that faith in science was never higher. Prospects for a scientific study of ethics emerge in this period—not without qualms from those "hard" scientists who would have preferred ethical issues to simply dissolve—but assuredly with a broad consensus in relativism of judgment, equality in views on human differences, and a keen sense that ethical concerns are empirically grounded, and without a need to resort to aprioristic assumptions either of a universalist or personalist variety. All of these experimental verities were buttressed by the social sciences: cultural differences in anthropology; democratic values in political theory; planning mechanisms to build a better economic order; safety nets to ensure equal outcomes and opportunity alike. In such a post-World War II world, fashioning a science of ethics seemed possible, if not exactly a piece of cake.

Forty years after works such as *Ethical Judgment*, such optimistic readings have been trimmed, if not exactly vanquished. From fundamentalist theologies at one end to deconstructionist culturologies at the other end, the faith in fashioning a scientific vision of ethics has been severely tested—not dissolved, but tested. If science in general is questioned, technology in the particular form of computerization of the globe has only accelerated the ideas of rationalization and standardization of knowledge. Ethical discourse now takes place in a high-tech environment not likely to be dissolved by appeals to teleological absolutes or solipsistic relativities. Technology itself has become an area of intense discourse: the place of privacy in a world of information overload; prospects for retaining copyright when the theft of intellectual property becomes so simple; and the place of knowledge itself in a world dominated by raw information.

The current outcry for moral certitude is in part a simplistic response to a world made complex by science and technology as such. Still, in part at least, the re-emergence of metaphysical certitude has been a Pyrrhic victory, based largely on a misreading of what people like Edel, operating within a naturalistic tradition, are trying to accomplish in the first place. Some of these concerns have struck deep. In an Edelian world, could one have refutation or simply argumentation and disputation? The broad brush that enshrined relativism in anthropologies in the pre-war epoch struck at the core of ethical theories in the post-war epoch. The assault from the cultural bastions were especially painful, since they came from the left side of the political fence from which people like Edel also emanated. Ethical theories grounded in relativist and naturalist assumptions, which began with a huge promise and sense of pride in the powers of human reason and social science, were placed on a back burner. They were not so much refuted by fanatics of the right or revolutionaries of the left, as by fears generated by prospects for nuclear warfare, revelations of the magnitude of genocides and the Holocaust, and the continuing scourge of inequalities in wealth and power. But the search for a scientific grounding to ethical theory does not go easily into the night. For whatever the fashions are of the political moment, the need to find a consensual basis for ethical belief and performance remains an everyday human imperative. But before we try to locate current contexts and what to expect in the proximate future, it is time to state plainly the contents of Edel's breakthrough monograph on *Science and the Structure of Ethics*.

At the end of the day, or better, the end of the century, this near-accidental publication of the *International Encyclopedia of Unified Science* represents a landmark in the development of a scientific study of moral behavior. The basis of this contention is Edel's singular capacity to move beyond oracular controversies of the good and the right in favor of a comparative, analytic, and functional account of how ethical perspectives and practices affect the content of moral discourse. In Edel's view, the structure of ethical behavior is defined by biological, psychological, social, and historical functions or orientations. Hence, a scientific account of ethics is possible since moral norms are themselves products of an existential field no less open to verifiable statements than other phases of human relations. He makes clear that adopting an existential perspectives framework (EP) does not require prior settlement of the free-will determinism controversy any more than the forging of political science depends for its sustenance upon the settlement of tensions between nations.

Edel's position is a serious attempt to move beyond the moral reductionism implicit in linguistic analysis as it existed at mid-century, and accepted by many self-described logical positivists with a minimum of discussion. In his view, the study of language divorced from meaning amounted to little more than a mandate ruling ethical conduct off-limits to the social scientist on emotivist, religious, or metaphysical grounds. Unfortunately, the logical disjunction between fact and value made earlier in the century by Max Weber and Emile Durkheim in response to intuitionist social theories may also operate to slow down attempts at establishing a science of ethics. What worked as a logical disjunctive became calcified into a metaphysical dualism. As a consequence, the place of ethics in everyday life failed to become the dominant theme in social research scholars might have expected or desired.

Edel's study thus emerged as a singular voice in a theoretical vacuum. The book is divided into four chapters: the first introduces us to the complex nature of the relation of science to ethics. The material here is quite condensed, and it might be advantageous to the reader to refer to Edel's earlier work, *Ethical Judgment*. Chapters 2 and 3, the core of the book, cover the theory of existential perspectives, and the methodological basis for a science of ethics. Here we are provided with an exposition of available (and inherited) ethical alternatives as well as the specific elements entering into ethical deliberations. The author's notion of existential perspectives is itself best understood as an effort to consider vantage points through which ethical judgments are framed, without becoming needlessly involved in strident defenses or critiques of these bastions.

In Edel's rich and varied taxonomies and paradigms, his scrupulous efforts to build an edifice only after firm foundations are provided become evident. In offering *ten* specifications for an existential perspective, *five* requisites for a scientific method applicable to ethics, *three* groups of moral subject matter, *five* types of linkage in establishing models for ethical discourse, *seventeen* forms of deliberation when adopting a course of ethical action, and *thirteen* meanings attached to the phrase "moral obligation," Edel recreates an Aristotelian style of philosophizing all but lost in the positivist canon. It may also have been a sly effort at attracting the support of these same positivists for whom precision became a value that at times extended beyond truth and meaning. The fourth and final section of the book is a plea for openness in discussing ethical beliefs and, no less, an offer of rich rewards for those patient enough to employ the findings of science in settling moral disputes. This, I take it, is what the author means by his "policy-making" approach to ethics.

This is not to say that questions did not remain at the close of *Science and the Structure of Ethics* that required further probes. Indeed, Edel devoted a great deal of energy beyond that volume in resolving several big issues. Within the confines of this brief introduction I should like to indicate several weak links within a firm expository chain. There is an ambiguous note struck in defining moral philosophy as asking the same kind of questions in the same sort of way as does a science of ethics. This is surely not held as true by most theorists in the field, nor for that matter is it so for many classical figures in philosophy. Is this something that Edel would *like* to see come about, or something that *must* happen in consequence of science and technology? I suspect that answers can come about only in relation to the development of the social sciences, in their capacity to examine the empirical correlates of ethical decision making.

However valiant the effort, I am not convinced that the notion of existential perspectives gets beyond, as it aims to do, Karl Mannheim's dilemmas in framing a relativistic science of politics in a universe of "total ideologies" and a "global relativism." What assurance can be given that there is not indeed a hidden or displaced existential perspective behind the facade of autonomy? And if this is so, how does Edel propose to get both scientific objectivity and existential perspectives squeezed out of the same general theoretical tube? Edel's later work on methods does come to terms with such issues. But that is food for a different meal.

There is a sublime contentment with relativism, drawn largely from an antiquated anthropology and sociology that is hard for me to share. Even if we separate functional problems from metaphysical issues, is there not a point reached where choice or selection between positions is required in advance of the evidence, that is, on essentially nonlogical grounds, such as the author's own faith in the curative powers of scientific endeavor? Is the science of ethics to rest on an exhaustive listing of EP's or is there a synthesizing principle involved? Edel's use of "policy decisions" as virtually synonymous with "virtue constellations" raises as many issues as it resolves. Are policy decisions to use scientific method in themselves free of, or determined by, moral boundaries? Aren't policy decisions often made on the basis of manipulative possibilities exclusively, without regard for moral dimensions as such? Would a Department of Defense approach to survival possibilities in a nuclear conflict become "ethical" by using the analytic tools offered in this study? At what point precisely do moral decisions coalesce with, or for that matter distinguish themselves from, practical decisions? Perhaps I have stated my queries more forcefully than the materials really warrant. It should be noted that these issues have come to the fore precisely as a consequence of Edel's lines of analysis.

As Edel well appreciates, and as the subtext of the book underscores, posing these kinds of questions presupposes an empirical attitude to ethics. And in compelling the community of scholars to redirect their energies along scientific channels, he has succeeded admirably in his main concern: establishing the parameters for an ethics of science no less than a science of ethics. The groundwork for an empirical study of moral behavior was provided in this brief monograph. What remains is the actual descriptive and comparative study of ethical considerations entering such phenomena as determination of wages and prices; the fixing of political consensus; the role of customs, habits, and traditions in moral decisions; the basis of moral responsibility by the mentally ill; actual moral ground rules in the growth of children; and a host of attendant considerations which are receiving far more attention among philosophers and social scientists at the close of the century than they did at mid-century. Edel's pioneering efforts helped bring research to the point where it was no longer necessary to *talk about* integrative levels, but actually makes possible the harder task of *integrating* levels and *breaking* barriers.

One would be quite mistaken to think that the sort of perspective enunciated by Edel was an isolated or unanswered *cri de couer*. Quite the contrary, a case can be made that the naturalistic tradition as exemplified by his *Science and the Structure of Ethics*, represents the spinal chord of moral theory in America throughout the twentieth century. True enough, it is a tradition which envisions creating a liberal society in largely egalitarian terms; but that has been the case for John Dewey through Richard Rorty. Edel stands at the mid-century point in what has become not simply an approach to the existential sources of ethics, but an effort to locate the guideposts needed to stimulate normative behavior based on a democratic consensus.

In a recent essay titled "Intellectuals and the Millennium" Rorty strikes a chord highly reminiscent of Edel's sense of historical perspective on the study of ethics. It is worth citing as a clear indicator that the democratic implications of naturalistic ethics remains very much the order of the day at century's end:

> We should not presume that there is a tight connection between the attainment of decency in human relations and the ascendancy of a particular world view. Nor should we imagine that any single European intellectual tradition—or Asian or African one for that matter—is clearly more favorable to bringing about a decent society than any other.

And in a particularly Edelian turn of the screw, Rorty adds that

> it may be that the differences between Confucianism and Christianity, or between scientific rationalism and post modernism, is just froth on the surface compared to the difference between a society in which there are untouchable castes and one in which there are not, or between a society where some people have thousands of times as much money as others and one where income differences are relatively slight. It may be that a decent society can be constructed without paying a great deal of attention to either religion or philosophy.

While it might well be that Edel would resist the implicit anti-intellectualism in the final caveat in Rorty's ruminations on the end of the epoch, his overall emphasis fits neatly with Edel's view of ethics as something grounded in human experience rather than mandated from divine assumption or mystic presumptions. So we return full circle to Edel's starting premises in an effort to construct a science of ethics. That Edel should repeatedly return to this centrist philosophic vision, tinged by a political liberalism that may or may not be a mandated requirement of that vision, it is hard for me to imagine a turning back from the hard lessons of the century—any more in ethical theory than in empirical research as such. We owe to Edel's examination of ethical doctrines an appreciation of the need to test and retest moral propositions in light of changing social and historical circumstances. This openness to experience informed by the lessons of physical science and social science now occupies a central place in our century's intellectual capital. *Science and the Structure of Ethics* helped to convert a common sentiment into a clear paradigm.

Science and the Structure of Ethics

Abraham Edel

I. The Nature and Complexity of the Problem

1. Issues in the "Relation of Science and Ethics"

Traditional views about the aloofness of ethics from science embody traditional conceptions of man and of science. Such slogans as "Science deals with the quantitative, not the qualitative," "Science deals with nature, not spirit," and "Science is theoretical, ethics is practical" give way before logical and mathematical analyses of order, the progress of psychology, the established importance of abstract theory in applied science. Contemporary reassessments in the philosophy of science as well as the tremendous advance of twentieth-century science call for recasting the problem of the relation of science and ethics in a fresh perspective.[1]

The complexity of the problem is evident from the several ways in which it can be formulated. A familiar way is purely *logical:* Can ethical propositions be deduced from scientific propositions? This leads to a theoretical impasse or, at best, a long detour. A more promising way is *logico-scientific:* for given meanings of 'science' and of 'ethics,' what patterns of relations (logical, psychological or sociological, pragmatic-instrumental, historical) can be envisaged and which can actually be found? How have these relation-patterns changed with shifting conceptions of the nature of science and ethics? A third formulation is necessitated by the discovery of changeable components in the patterns: How far is the relation of science and ethics an *evaluative problem,* requiring policy determination?

In such restructuring of the problem we must distinguish *the place of scientific results in ethical theory, the role of scientific method in ethical theory,* and *the impact of the scientific temper in ethical theory.*

1

2. The Place of Scientific Results in Ethical Theory

It is generally agreed that scientific results furnish the best way of improving means-judgments in ethics; but there is still a strong tendency to concentrate solely on this function. Accordingly, little has been done toward analyzing other points at which scientific results enter into ethical theories. One consequence is that there is no ready way to study the shifts that may be required in an ethical theory because of fresh advances in psychology or the social sciences or of assessing the extent of such relations.

A major source of difficulty in determining the relations of scientific results to ethics stems from the fringe-boundaries of the concept. In the physical sciences we can specify quite readily what are 'scientific results.' In the psychological and social fields, some would maintain that we have few results, but only areas of promise—models or schemata rather than specific established theories. It is not our purpose here to appraise the state of development of the several areas. Many questions for which there are not yet answers have moved sufficiently far into the domain of science to be regarded as scientific questions. For example, the psychology of perception has results; the psychology of thinking is more disputable; the relation of willing and thinking shares in the difficulties of the latter. We may distinguish scientific questions with established results as answers, scientific questions with competing theories as proposed answers, scientific questions with even the direction of answers unclear. In some cases it may not even be clear to which of the existent sciences the enterprise of answering is to be assigned.

On the other hand, the concept of 'scientific results' should not be limited to general theories. It covers also 'scientific findings,' where what is discovered is the existence of fresh specimens—a hitherto unknown metal, or species of insects, or kinship pattern, or a different moral code. Some theorists have attempted to draw sharp distinctions between fact-finding history and generalizing science. For some purposes sharp lines may be useful if achievable; but for considering the place of scientific results in ethics, particular findings of history or anthropology that throw any light or suggest any lessons come well within the scope of 'scientific results.'

What kinds of scientific results can we expect to have some relevance for ethics? In a broad sense, any that add to our picture of the world in which man lives and acts, that throw light upon the nature of man and his capacities, social relations, and experiences. Biology furnishes a picture of the constitution of the human animal, his place in evolutionary development, his instinctual equipment and basic drives, his interaction with the environment. Psychology explores many human powers basic to ethics. There are studies of pleasure and pain, of feelings and emotions, volition and inner conflict, basic needs, mechanisms of defense and conditions of insight, development of character and personality, varieties and conditions of moral feelings and their phenomenological description, psychiatric materials on amorality and on moral rigidity, patterns and qualities of interpersonal relationships, and so on. The social sciences, whether general as anthropology and sociology, or oriented to specific phases of institutional life as economics and political science, or specifically problem-centered as education or penology, offer a constantly growing fund of information and suggested generalization. They depict the basic and recurrent problems of societies, the variety of patterns in which men have tried to solve them, the kinds of group organizations men have elaborated, the extent of their success or failure, conditions and problems of stability and change, the human costs of social organizations of different types, and so forth. Included within their scope is the social history of morality itself and of ethical theorizing. Similarly, history insofar as it goes beyond the reconstruction of particular events furnishes some accounts or suggestions about the growth of basic aspirations and strivings, about the emergence and career of various ideals, about major instrumentalities and the impact of different kinds of problems and events on human hopes, and it provides some materials at least for speculation on what under what conditions is likely to be unavoidable, and what room there is for men's reconstructive efforts.[2] These are the kinds of materials with whose impact in ethical theory we are concerned.

The extent of such scientific results—the reach of science in dealing with human phenomena—depends on whether the state of the field is such as to set limits to applicability of scientific method. The field must be sufficiently determinate to admit of isolable and recurrent elements, and this means that the phenomena must not be unmanageably complex and unmanageably unstable. For general reference, let us speak of the *indeterminacy of a field* as constituted by its *field instability* and *field complexity*. Some of the ancients thought that the physical world was in such constant flux that no stable knowledge of it was attainable. (Aristotle tells us that this view of Heraclitus led Plato to turn to the purely mathematical domain as the prototype of knowledge.) Many of the moderns believe that the human field is characterized by a major indeterminacy of one sort or another. Theories about these are equivalent to theories about the presumed unavoidable limits that the sciences of man will reach in their development. Part of the task of tracing the place of scientific results within ethical theory is to render explicit such field assumptions operating as the basis of judgments about the nature of ethics, so that they can be assessed in terms of the scientific evidence.

One small part of the relevant area of scientific results is the scientific study of directly moral phenomena. It is obvious, though insufficiently stressed,[8] that there are no restrictions on the scientific study of these phenomena. Whatever be the interpretation within ethical theory of moral utterances such as 'I ought to do X' or 'Y is good,' there is for each the corresponding descriptive statement that a person is having a certain kind of experience. Whether the phenomenon it describes can be *successfully* studied scientifically depends in turn on the state of the field.

3. The Role of Scientific Method in Ethical Theory

How far can the methods of defining, locating phenomena, isolating data, discriminating observations, carrying out experiments or carefully controlled observations, dealing with alternative hypotheses, utilizing logical techniques of classifying and systematizing, verifying and establishing a body of reliable knowledge in general form, and so forth, be applied in ethics? How far is ethics in fact in a prescientific stage, but capable of being developed? Or is this a misguided illusion? What are the necessary conditions for such a development and how far does or can ethics satisfy them? On many of these issues contemporary moral philosophers are seriously divided.

There are general attempts to argue that if scientific method proved applicable *within ethics*, that is, if utterances of the form 'X is good' and 'I ought to do Y' could be established scientifically, the body of such assertions would constitute a science and not ethics. A theoretical science, it is argued, acquires a body of well-demarcated phenomena and looks for laws on the basis of which there can be prediction, and theories that will systematize the laws. A practical science aims not to explain but to secure certain results in production or conduct. Ethics is prescriptive; it tells a man what he should do, not what he will do; it is therefore practical. But, of course, engineering, medicine, psychoanalytic therapy, education, are all practical disciplines, and yet they are markedly different in the extent to which they are scientific. Engineering is predominantly scientific, medicine in many parts. Education is scarcely so, and psychoanalytic practitioners sometimes stoutly defend the view that theirs is an art rather than a science. But educational theory as part of the study of cultural transmission and psychoanalytic psychology as part of the psychology of personality are clearly scientific in intent if not in accomplishment. Even if they were so in accomplishment, it would still be a distinct question how far the practical discipline could or could not apply their results and utilize scientific methods.

In a similar vein it is sometimes argued that if 'theory' means the same in 'an ethical theory' and 'a physical theory,' then an ethical theory is a theory *about* morality rather than within the ethical field; it is a scientific theory rather than a formulation yielding moral guidance. Certainly, a theory about fishing is not fishing, and a theory about advertising is not advertising, any more than a theory about motion is itself a motion. But a theory about fishing may be the most practical way to help organize successful fishing, and a theory about advertising may help restructure advertising activities. Similarly, a theory about morality may be the only way to *understand* morality as well as the most effective guide.

What is really involved in these arguments is whether ethics requires a "logic of the will" distinctive in type from the procedures of scientific inquiry. This is a legitimate problem, but the answer does not follow from the mere acceptance of practical aims in the discipline. It is more helpful to pinpoint the differences that are presumed to exist in ways of treating concepts, in expectations of the possibility or impossibility of working out verification procedures, and so forth.

The problem of the role of scientific method within ethical theory is not coincident with the place of scientific results within ethical theory. Suppose it were a discovery of psychology that man is so irrational that on basic questions of human purpose men were incapable of thinking logically or rationally. Then obviously, since using scientific method involves reflecting rationally, men could not use scientific method in moral processes. It would thus be a scientific result that scientific method had no role in ethics! (In fact, irrationalist claims usually embody some assumptions about the nature of the human will which yield this type of result.) However, this example further suggests that, while the two questions are different, their answers are interrelated. If the human field is sufficiently determinate and scientific results have a constitutive place in ethics, then the wider use of scientific method within ethical theory may be possible. Whether to take advantage of this possibility would be a decision of methodological policy. To see what considerations enter into such a decision requires a detailed examination of the structure of an ethical theory and the nature of its conceptual and methodological apparatus.

4. The Impact of the Scientific Temper in Ethical Theory

The contemporary scene has witnessed increased impact of a reflective analytical-empirical outlook on ethical theorizing. Calls for clarity, for the separation of analytic from empirical issues, and of decisional from both, are frequent. Concern among philosophers with the logic of ethics and with the language of ethics, and the rise among anthropologists of empirical value studies, further reflect the scientific trend.

In the case of the scientific temper, the evaluative character of the relation of science and ethics emerges quite clearly, for there are alternatives. There is also the mystical temper, or the activist temper. Which to employ is a policy decision. Even more, it is clearly an ethical decision; for a choice of temper is in effect a selection of a virtue-set. Compare this with the question of the place of scientific results in ethics. Here there may be little choice. If there really are gaps in any ethical theory which have to be filled in somehow by a conception of the self or a conception of the relation of intellect and will, then the only choice may be whether to use outworn scientific results or contemporary scientific results. (Of course, there is always the alternative of using myth or ideology or of embodying partial results geared to produce a particular ethical outcome at all costs.)

Policy decisions in ethical theory share with all basic policy decisions the problem of vindication—a complex but not insuperable problem. The attempt to go as far as possible toward the use of science in ethics is part of a unified philosophy of man and his world, which presents a promising present policy. The bases for policy decision in dealing with scientific results and scientific method in ethical theory will be considered in the body of this work. The question of evaluating the virtues which constitute the scientific temper is a particular ethical problem which lies beyond the scope of this general study. It will be touched on briefly at the end.

5. Moralities

The history of writings on ethics shows that ethical theories are reflective attempts to understand and help guide morality. Understanding often involves providing principles for systematizing and interpreting; guidance may take the form of adding confidence and furnishing basic direction through justifications or offering organized programs of modification. Ethics is a reflective enterprise; moralities (frequently called 'moral codes') have a more overtly regulative character. In addition, actual moral codes and propounded ethical theories have had varying relations and sometimes relatively independent historical careers. On the other hand, the intimacy of their relation is witnessed by the fact that the answer to the question "What is a morality?" is a fundamental part of an ethical theory.

In the array of propounded ethical theories there are many different answers to this latter question. Some locate their phenomena as men's moral judgments which they construe as individual acts and examine psychologically or phenomenologically; here are to be found theories of moral requiredness, indefinable moral qualities, morality as some dimension of interest or appetition, and so on. Others, starting equally with individual moral judgments, give them a linguistic cast; moral statements are identified by the presence of moral terms ('ought,' 'good,' 'duty,' etc.), and ethical theory takes the form of logical analysis. Others, still focusing on individual discourse and looking for the "normative force" of ethical utterances, find their basic phenomena in specific functions of the utterance —expressive, hortatory, commendatory, etc. A rather different initial approach is found in those who approach morality as a sociocultural phenomenon and locate their initial data in group processes and institutional forms.

A full delineation of a morality has to include all the features which different views invoke for varying theoretical purposes. A morality embraces many qualities and processes: rules enjoining or forbidding selected types of action, selected character-traits cultivated or avoided, selected patterns of goals and means. These in turn are organized in the behavior and consciousness of people in a variety of ways—a conception of the community concerned and a responsible person, a set of ethical terms and some scheme of systematization of discourse involving them, some pattern of justification processes, some types of sanctions, some selection from the range of human feelings tied in with these regulative procedures. Thus typical Western morality has its code of ten commandments; its virtues and vices; its goods of happiness or salvation or success; its universalistic moral community in which "the dignity of man" is taken to insure that everyone counts; its notion of individual responsibility; its language of 'ought,' 'right,' 'wrong,' 'good,' etc.; its religious or utilitarian justifications; its heaven-and-hell or internal-conscience sanctions; its concentration on guilt and shame. The ideal of scientific description of moralities would be a moral map of the globe over the history of man, just as one could have a linguistic map or a map of religions. This presupposes coordinates of description, principles of classification, and some determination of configuration-types.[4]

6. Methodological Approaches in Ethics

If the role of science in ethics is to be fully explored, it is important to include all the methodological approaches operative in ethical theory today, dispersed though they may be. Four standpoints are suggested as pertinent.[5]

(1) The analytic approach provides reflective examination of the conceptual apparatus of ethics—the meanings, uses, and relations of ethical terms; formation rules and modes of reasoning; ways of justification; and so forth. It helps distinguish stipulative, factual, and purposive components. Its outcome consists in conceptual schemes and methodological guides.

(2) Descriptive approaches focus on the experiences and processes that are identified either as moral or as relevant to the solution of moral questions. There are various descriptive modes —the familiar behavioral and introspective modes, phenomenological description reporting qualities and relations in the field of awareness, sociocultural description adapted to group phenomena, historical description concentrating on succession of phenomena over large time-spans.

(3) Causal-explanatory approaches look for the conditions and contexts of the occurrence of moral phenomena, experiences, processes. In whatever terms these phenomena be described, there is always the question of relating them, as they occur, to other phases of the life and internal economy of the individual and to the other phases of operation of the society at a given point of development of its history and culture.

(4) Evaluative approaches involve the employment of standards to assess whatever is the focus of attention. They are operative in decision, in the justification or criticism of decision, in the establishment of principles and policies, in the formation and reiteration of commitments, in the development, stabilization, and alteration of standards themselves.

Each of these standpoints brings to bear methods increasingly refined in contemporary logic, science, and human experience. The lessons of modern logic, mathematics, and linguistic inquiry give force to analytic sifting in ethics; for example, the traditional overready resort to moral "axioms" rested on appeal to arithmetic and Euclid, and must reckon afresh with contemporary interpretations of mathematics. Modes of description and observation have been refined in the controversies of psychological schools and in both social science approaches and phenomenological philosophies. Causal investigation opens the door to what can be learned about the career of morality and ethical ideas in the light of various psychological, cultural, social, and historical schools of thought. Evaluative processes in other fields—in law and education, as well as economics and politics, even procedures of standard-development in pure and applied physical science— help throw light on comparable moral processes.

In any particular inquiry in ethical theory, care must be taken to recognize which enterprise is being carried on at what point and in what respect. Take, for example, the ethical problem of *obligation*. It is an analytic problem to determine the possible meanings and various uses of 'ought,' how far its use differs in such expressions as 'I ought' and 'He ought,' whether it is meaningful to say that something 'ought to be' or only that it 'ought to be done,' whether 'ought' implies 'can' and in what senses, and so on. It is a descriptive problem to give a clear representation of the feeling of obligation, or the pangs of conscience, or the occurrence of requiredness in the field of awareness, or the type of interpersonal relation that is designated by such phrases as 'recognizing a claim by another upon oneself'—all this on the assumption that we have in these various cases some definite experience we are referring to. It is a causal-explanatory problem to discover the conditions (whether they be psychological or sociocultural) under which we find a harsh conscience or peremptory obligation-feelings, or under which the obligation is felt as imposed from without or issuing from one's self. It is an evaluative problem to determine whether promise-keeping is a moderately weighty or an absolute commitment or whether, if possible, men should cultivate humanistic rather than authoritarian consciences.[6] Confusions clearly arise if one simply asks "What is obligation?" without distinguishing the type of inquiry intended.

7. The Structure of an Ethical Theory

What are we to understand by (i) *an ethical theory,* (ii) *the structure of a given ethical theory,* and (iii) *the structure of ethics?*

(i) An ethical theory provides, broadly speaking, an analysis of the basic concepts and methods of a morality, a descriptive account of the types of phenomena involved, an explanation of the relations of the morality to the fuller context of human life, and procedures for using the morality in deciding and evaluating. There have obviously been many ethical theories (e.g., Platonic, Stoic, Kantian, Utilitarian). In the textual and historical materials from which each of these is extracted, we can distinguish between the morality as a pattern of virtues, rules, and social policies, sanctions and feelings, and so on, and the ethical theory of the human makeup and its strivings, the basic order and its necessities, the conceptual analyses of 'good' and 'obligation' which are offered to expound and support the morality.

(ii) To speak of 'the structure of a given ethical theory' is to call attention to the kinds of problems to which the particular ethical theory is addressing itself in its attempt to understand and guide the morality with which it is associated. On the whole, the structure of traditional and contemporary ethical theories becomes clear if we see their assumptions about the human field in its world background, the concepts and methods they employ, their account of the way in which standards are established and decisions made, their specific interpretations of freedom and responsibility.

(iii) To speak of 'the structure of ethics' is to go beyond the particular notion of 'the structure of a given ethical theory.' It assumes a certain community of enterprise in different ethical theories—that they are addressing themselves to the same central problems however different their solutions. This need not be an unexamined assumption; it is possible to render the basic problems explicit and to study the variety of their formulation and the limits beyond which we would cease to be dealing with a common field.

If the problems are comparable, it becomes possible to suggest co-ordinates for the mapping of ethical theories, to look for possibly invariant features, and to analyze the role that different parts of theories play in the functioning of the theories themselves. To explore the relation of science and the structure of ethics is to go along a major avenue of such comparative judgments.

II. The Theory of Existential Perspectives

8. The Concept of an Existential Perspective (EP)

A striking fact in the comparative study of ethical theories is their diversity. There are individual and social types, otherworldly and this-worldly, inner-oriented and outer-oriented, intuitive and empirical, rigorous and genial, conformity-directed and welfare-oriented, and so on.[7] The issues among them include historical factors and valuational conflicts. Our concern here is to discover how far the differences are due to different answers to scientific questions, operative within the ethical theories. Our heuristic principle is that this factor plays a major role in the inner structure of ethical theories, that it is sound policy to render it systematically explicit, and that a sizable part of the evaluation of ethical theories (and therefore of the task of reconstruction in ethical theory) centers about it.

From the point of view of this inquiry, ethical theories divide into three kinds: those that specify a basis in scientific results, those that specify a trans-scientific basis (usually theological or metaphysical), and those that ultimately deny the relevance of any existential basis at all to value theory. Our procedure of inquiry will be as follows:

a) To construct a general or abstract concept—the *existential perspective of an ethical theory,* hereafter abbreviatel to EP—to refer to assumptions about the world and human nature, images of man, etc., operative in an ethical theory. Such a general concept leaves open the types of specialization it will take in different ethical theories—e.g., metaphysical, psychological, historical. As here used, the term 'existential' refers to a way of viewing *existence.* It is to be distinguished from the term 'existentialist,' which has come to designate one special theory about existence. An existentialist EP would thus be one type of EP.[8]

b) To explore the *overtly scientific* types of EP and suggest the extent to which differences in the ethical theories that have such EP's stem from their reliance on different sciences or selective emphases in the psychological and social sciences or different stages in the development of these sciences.

c) To explore *theological and metaphysical* types of EP to see how far they embody different answers to scientific questions, and what role these play in the ethical theories that have such EP's.

d) To criticize the claims of "purist" ethical theories that disparage the role of existential assumptions in the inner workings of ethical theory and to suggest that they have a hidden or "displaced" EP behind the façade of autonomy. This will be called a *transcendence* EP, and the same kinds of questions can be asked about it as about the theological and metaphysical types.

e) To suggest criteria for the evaluation of EP's as part of the general task of securing a more adequate ethical theory in the modern world.

The EP of a given ethical theory is its view of the world and its properties, man's nature and condition, insofar as these enter into its understanding of moral processes and moral judgments. In looking for the EP of a given ethical theory, we could ask how far it employs:

A. a particular view of the world and its constituents
   a particular theory or model of its processes and mechanisms
B. a particular view of the nature of man and his dominant aims
   a particular reference to unavoidables in life and action (birth, matura-
      tion, reproduction, aging, death)
   a particular theory of men's faculties—intellectual, emotional, practical—
      and their relations and an image of the self
C. a particular image of community—its nature, bonds, extent
D. a particular consequent view of the degree of knowability of the world,
      human life, community, and their processes
   a particular view of determinateness and indeterminateness in their pat-
      terning
   a particular expectation of dominant dangers (e.g., illness, human ag-
      gression, etc.)
   a particular assessment of control-possibilities, including (where it
      exists) any estimate of morality itself as a control-instrument.

In recent sociological and anthropological theory there has been some movement toward a configurational concept in describing a group's values. For example, Kluckhohn adapts the concept of a 'value-orientation' to this purpose and defines it as "a generalized and organized conception, in-fluencing behavior, of nature, of man's place in it, of man's relation to man, and of the desirable and nondesirable as they relate to man-environment and interhuman relations."[9] Redfield explores "world views" as "the way a people characteristically look outward upon the universe," "the structure of things as man is aware of them."[10] Others explore "ideologies" and their components, or varieties of philosophical assumptions.[11]

Such materials are very suggestive for EP dimensions likely to be over-looked if we concentrate on the Western ethical tradition. The EP con-cept, however, is a narrower one in two respects. It is concerned with the stage-setting for morality, which need not coincide with *total* world-outlook, and it aims at winnowing out analytically the *existential* aspects, no matter how fused they may be in the people's outlook with value-components.

A useful device for working out EP dimensions is to employ a theatrical metaphor: the EP is the *stage-setting* for the performance embodied in the ethical processes.[12] Thus we may think of the scenic *space* of an ethical theory: moral problems are set for a man as an issue of what is going on inside himself, between himself and other individuals, between himself and a group, between himself and the cosmos. Or, again, it may set them as primarily group problems or whole-world problems. Similarly, the *time* may be here-and-now (focusing on the quality of present experience), the past (e.g., moral duties construed as debt to one's ancestors), the future (governed by distant goals), temporally indeterminate or "aorist" (concerned with the eternal).

What of the *dramatis personae* for the performance? Some EP's include gods as well as men; some have men alone. Some have whole groups, cast as nations, species, or classes. Others use the central dramatic figure of the individual man. Where the stage is further limited, the actor may be some part of man— a soul, a rational element, a superego, a host of desires, or instincts, drives, and needs.

As for *typical lines of action,* the script often assigns well-demarcated roles. For example, in religious stage-settings, the Greek gods act like exalted humans with all their passions and conflicts, the God of the Decalogue like an exalted father; in West Africa the stage contains both Ancestral Spirits concerned with the behavior of their lineal descendants and individualized Fates.[13] Nations or peoples unfold destinies or accumulate power, species struggle for survival, classes for domination. Souls seek salvation or beatitude in relation to a divine figure; sometimes they are torn in the struggle of divine and satanic beings. Rational elements do everything, from maintaining Kantian consistency or engaging in Benthamite calculation to restraining strong drives in Pueblo Indian conformity. Superegos specialize in inflicting guilt-feelings or parade in judicial robes. Drives press on, needs seek satisfaction; hosts of desires usually stumble over one another until conditioned into harmonious equilibria. Some plots require richly decked stage-settings; some, very bare ones.

Similar suggestions can come from the typical environment of the action: the non-human world as means to be manipulated, or as possessed of compelling power, as forcing certain unavoidables or presenting dangers. The stage may be brightly lit (intelligible) or obscure and murky (inscrutable). What is its *steadiness*—absolute fixity or near-chaos (in which a good Stoic is ready to face anything)? What is the degree of *determinacy in plot structure*—a fixed plan with no alternatives, or plenty of room for choice? What is the *central focus* of the characters and their action? Two stage-settings may both include the individual and the group. But in one the group is external to the individual, an instrumental help or hindrance to his goals; in the other, the individual is essentially a group being. Both differ from an EP that leaves the individual out and speaks only of group survival and expansion. (Psychologically, the question of central focus may be seen here as one of self-involvement.)

A concluding warning: we should never underestimate the ingenuity of stage-designers (or the complexities of human life): Hegelian ethics uses a simultaneous two-scene setting—the world process in one, the individual consciousness with its pressing passions in the other—and central focus is on the cunning of reason which builds the action of the former on the materials of the latter.

9. Role of Existential Perspective within an Ethical Theory

How far EP answers are answers to scientific questions, and how influential they are in ethical theories, is a basic question in the rest of this work. By, way of anticipation, let us illustrate the range of influence in one type of theory with an overtly scientific EP—the Utilitarianism of Bentham and Mill.[14] The scenic space of its EP is this world of men; its time is the period of worldly life. Its dramatis personae are individuals treated as units, not communities or nations. Its typical lines of action are pursuit of pleasure and avoidance of pain; so strongly is this stated that we are likely to overlook the concomitant activity of forming associations according to psychological laws. Causality is taken for granted in the operations of man and nature; men age and die; there is always a strongly operative background of other people; individuals are very different in their sensibilities and the circumstances that determine them; religious forces enter the picture only through beliefs in them which have an influence on conduct. The stage is fairly luminous: men can reason and calculate; they need not simply hold on to blind custom. There is sufficient steadiness and determinacy so that men can successfully plan and generalize and apply the lessons of their experience; what is more, one can count on their following certain lines once they fully realize that it means increased happiness and once they have formed the appropriate associations. The central focus in the theory is social, in the sense that there is a primary interest in the shape of social forms and institutions; at times there is a double focus, with shuttling between individual and society.

How does this particular EP tie in with the remaining constituents of the structure of Utilitarian ethics? The reference is not to causal determination but to relations varying from logical implication through instrumental and functional gearings to different degrees of parallelism or similarity. A number of tie-ups stand out quite clearly. The picture of men's basic strivings is directly incorporated into the definitions of ethical terms: 'good' is analyzed as pleasure or what produces a pleasure surplus, and 'ought' in terms of the line of action conforming to the greatest happiness principle, and comparably for other ethical terms. The syntax of ethical expressions follows consistently; for example, the subject term in the expression 'X is good' in its primary use is an experience term; comparative terms (e.g., 'better') are analyzed in terms of degree or quantity of pleasure, as in the familiar felicific calculus; and so on. The assumed rational abilities of men and the luminosity and determinateness of the stage make it possible to ascribe an inductive methodology for ethics, with lessons of experience about what is productive of the greatest happiness. Operative indices for applying moral rules are linked to assumptions in the stage-setting; for example, since sensibilities differ, every man is his own best judge of what pleases him. Techniques for social stabilization of the morality similarly are linked to assumptions about the way in which men are influenced—not merely the detailed theory of sanctions as specific forms of pleasure and pain inducement but the concentration on such fields as economic relations, law, education, and formation of social habit rather than merely on moral discourse, as some ethical theories are prone to do. These areas of application also express the central focus, with its aim to provide standards for social judgment and social reform. We need not enter into the content-values of the theory; some, such as the high appreciation of rational activity as a pleasure, certainly parallel the place they occupy in the EP. But we may complete this brief survey by noting that the very nature and tasks of ethical theorizing are worked out in a way congruent with the EP in its several dimensions—ethical theorizing is itself a sharpening of the tools for calculation or reckoning so that the pursuit of happiness will be more rather than less effective, far-reaching in plan rather than blind or tradition-bound.

Just as in Utilitarian ethics views of men's world, nature, motivations, interactions, guide and help shape formal and methodological elements, so too in other ethical theories we can trace a similar influence, though the precise pattern of influence varies with the content.

10. Overtly Scientific Existential Perspectives: Physical and Biological

Overtly scientific EP's may be extracted from ethical theories that have actually been propounded and also by looking to the specific sciences to see what their offerings are for ethical theory.

Physical stage settings have been offered in principle, but scarcely more. A Democritus or a Hobbes may construe man as a bundle of particles or physical motions, but he is soon given a physiological and even a psychological costume before he begins to interact with his fellows. Sometimes the physical description provides a model for the transfer of some properties to human action (e.g., self-maintenance, self-preservation), or some of the "laws" of human behavior are cast in a mold analogous to the laws of physics. Physical EP's therefore appear usually as physical-biological or physical-psychological.

Yet it would be a mistake to ignore the power exercised even by the promissory notes of this EP. Its ethical potential is sometimes far-reaching. It eliminates interpretations of duty or obligation in terms of gods or souls. It points a direction for the reduction of ethical concepts—for example, of good to feeling pleased or goal-seeking, and of conscience to a special form of fear, and these in turn to an internal movement or a drop in tension and to a disorganization state or special tensions, and so on into chemical-electrical processes. The causal emphasis in the physical stage-setting usually means an interpretation of will-acts as resultant phenomena occurring in a lawlike manner, as in Hobbes's description of will as "the last appetite, or aversion, immediately adhering to the action, or to the omission thereof."[15] Nor are we dealing merely with philosophical history. Physical stage-settings have increased significance today in relation to the live interest in the mechanism of "thinking-machines" and the attempts to work out concepts of teleological mechanisms. Similarly, construing the body as a stabilized machine with determinate modes of action becomes readily translatable into a system of inherent or unalterable tendencies seeking fulfilment, and so into an egoistic individualism.

Biological stage-settings derive their concepts from several sources. One is the study of the individual organism as it develops and functions. The focus of such an EP is on intraorganic units or on individuals. Many an ethical stage is set with 'impulses' or more determinate 'instincts' made into stable unquestioned starting-points to which ethical activity is held responsible.

Other biological EP's focus on groups—small, or whole populations, or even humanity as a single global population—and devote special attention to properties characterizing the evolution of the species or the biological history of populations. Sometimes we find a single strand selected for a central role, such as the struggle for existence in nineteenth-century survival-of-the-fittest ethics, or the phenomenon of mutual aid, as in Kropotkin's ethics.[16] Sometimes we find an organization property, such as adaptive harmony or adjustment, or maintenance of a special kind of equilibrium providing key slogans for the ethical processes enacted on their stage. Social-organism theories of ethics exploit the analogy and call on biology to provide the picture of how men became so integrated in their human social reactions;[17] or the homeostatic model is transferred to social organization and to ethical regulation. Sometimes attention is fastened on a developmental thread or a persistent trend, such as increase of population, spread of the area of human control, development of larger co-operative aggregates, and so forth; these tend to be cast as criteria of "progress."

A third type of biological EP sets a full-scale evolutionary stage. While its concepts are evolutionary, its materials usually go far beyond biology. Herbert Spencer's picture of evolution in all fields from chaotic homogeneity to organized heterogeneity is a classic illustration. Such stage-settings often deal consciously with the concepts of ethics and the structure of ethical theory, as Spencer sought to do, and attempt to incorporate changing elements within ethics into their account of evolution itself. The strongly ideological character of Spencer's system has obscured its theoretical scope.[18]

11. Overtly Scientific Existential Perspectives: Psychological

Psychological EP's have been perhaps most common in modern ethical theories. Their philosophical articulation has tended to follow the lines of the development of psychological theory itself.

The classical *introspective* type is most clearly seen in the theoretical formulations of the British empirical tradition from Hume to J. S. Mill. The building-blocks of ethics are feelings or sentiments, introspectively discernible states of consciousness —whether pleasures and pains, acts of sympathy, driving passions, or more deliberative calculations expressive of self-love and fear.

To locate the elementary units and exhibit the mode of combination—for example, to pin down sympathy and show how it is built up into feelings of duty, as Adam Smith does,[19] or to pinpoint pleasures and pains and show modes of measurement of large-scale "lots" in terms of their components, as Bentham does[20]—constitutes a great part of the task of such ethical inquiry. A sizable part of contemporary empirical ethics retains this tradition. Bertrand Russell's stage-setting is the matrix of desires with a special focus on the conflict of impulse and intelligence;[21] Schlick's focus is on the pleasantness or unpleasantness of ideas in the motivation of willing;[22] and a great part of general value theory is cast in affective terms of desire and feeling (described as states of consciousness) as the essence of the value phenomenon.

With the rise of *behaviorist* tendencies and the revolt against introspection—especially in ethics because of its dualistic associations—an emphasis on observable behavior replaced the compounding of states of consciousness. The psychological stage-setting was conceived to be an extension of the biological picture of the organism and its adaptive responses. Directional concepts could not be wholly discarded but could be seen as selective tendencies, propensities to action, and the like, in the human organism.

Ralph Barton Perry's *General Theory of Value* (1926) exhibits clearly this type of EP. The basic concept of *interest* in terms of which Perry identifies value is given a biological basis and a psychological articulation: "Interested or purposive action is action adopted because the anticipatory responses which it arouses coincide with the unfulfilled or implicit phase of a governing propensity."[23] Modes of interest, the role of cognition, types of integration, are studied, but this does not carry us on to a social setting of the stage; on the contrary, Perry says, "we are concerned only with society in so far as it is a composition of subjects who interact *interestedly*, or are integrated in and through their interests."[24] Two influences in the growth of behaviorist psychology enrich this type of stage-setting. The first is the liberalization of behaviorism in such work as E. C. Tolman's,[25] in which 'goal-seeking behavior' comes into its own, with rigorously controlled criteria for identifying the phenomenon. The corresponding ethical stage-setting is still individual-psychological, with distinctively human aspects found in the proliferation and organization of the scheme of purposes in group environments and in the role of cognition mediating the rise and fulfilment of goal-seeking. The second influence is the impact of Freudian depth and basic need emphasis, which led the behavioral stage-settings to give a more explicit place to need-concepts, without however surrendering the insistence on observable verifiability. The difference that these two influences make in the refashioning of the psychological stage-setting is apparent if we compare Perry's early work with the mapping of the structure of appetition and aversion that we find in Stephen Pepper's recent *Sources of Value*.[26]

Freudian psychology and its derivative schools have provided materials for a whole class of stage-settings that have been explicitly applied to ethics. The analyses of conscience, character-development, internal conflict, have an obvious bearing on questions of obligation, virtue, and freedom in ethical theory. More sharply, the Freudian stage is specified by the career of the instincts in the development of the individual, the differentiation of the id in the rise of the ego and the superego, their modes of interaction and conflict, the types of sublimation or neurotic distortion or breakdown that ensues—in general, a theory of the maturation of personality. In the maturation process a pleasure principle is modified by a reality principle which admits of deferment in gratification; conscience is depicted as a specialized development of anxiety in intrafamilial reactions of the oedipal phase; and morality is generally construed as a repressive mechanism against aggressive tendencies. The Freudian stage reaches back into a biological stratum, at least in Freud's later biological derivation of the Life instinct and the Death instinct. It reaches out toward a social structuring by seeking to find the "cement" in interpersonal relations that make wider group association possible, and it assumes historical contours by picturing the growth of civilization as the increase of group affiliative extent, resting on a constantly greater repression.[27]

Variant theories of personality within the broad psychoanalytic outlook produce many readjustments in the ethical stage. Conceptions of an independent conflict-free source of ego development rooted in processes of perception and motion[28] yield a less repressive picture of the tasks of morality and a greater role for a positive picture of the self. Theories that tie aggressiveness to frustration rather than an instinctive base yield an EP with more room for action.[29] There are variants in the specific steps of the developmental process, the outlines of character-formation, the role assigned to cultural influences, and so on, all of which have different ethical potentials and some of which have been applied to ethical theory.[30] Some psychoanalytical writers, such as Sullivan, shift the whole focus of the inquiry to an emphasis on interpersonal relations.[31] As an EP this ceases to be individualistic-psychological and becomes at least two-person psychological; its ethical import, though not as yet analyzed, would seem to be considerable and not unlike the broadening in those psychological schools that stress a constitutive role for cultural elements. Finally, whatever the variety of psychoanalytical approaches, there are some findings with considerable ethical impact—a conception of basic human needs as scientifically discoverable and as demanding a place in any EP, or a recognition that there are techniques of systematic self-deception, which challenges the ultimacy of individual introspective value reports and renders them in principle partially corrigible.

Phenomenological psychological approaches are the modern heirs of the older scrutiny of consciousness and the moral sentiments. Their scientific development in contemporary Gestalt theory embodies a revolt against the behaviorist neglect of the "meaningful" aspects of experience and yet an unreadiness to equate these with the older introspective picture of states of consciousness. In the writings of some psychologists and some philosophers[32] the stage is set with men as beings who have a level of experience with qualities of its own as moral, and who find in the field of this experience such contents as "meanings" and "values," and who are capable of inspecting the field and enunciating relations. There is the confidence that general and stable results can be achieved by phenomenological inspection of moral experience just as laws have been found for the visual field and other areas of perception phenomena. There is no attempt to restrict causal research into men as bearers of such fields, but this is construed as an external enterprise, involving the correlation of physical or physiological variables with field variables.[33]

There are, of course, serious attempts today to develop a unified psychological model of man in which phenomenological and behavioral methods will be employed together with conceptions of underlying needs and developmental processes. The success of such attempts would have coresponding effects on the available EP's of the psychological types.

22

12. Overtly Scientific Existential Perspectives: Sociocultural and Sociohistorical

Many ethical theories that present no systematic picture of their existential assumptions are prone to rely on an interpersonal or general sociocultural account of the context in which ethical problems arise. For example, emotive theory focuses on disagreement in attitude, which involves at least a two-person group, in setting up its fundamental paradigms for interpreting ethical language.[34] The same is true of prescriptive types, for which the distinctive function of ethics—commending, advising, persuading, etc.—is clearly interpersonal. Where a sociocultural EP is explicitly advocated, it represents a theoretical conclusion that morality is in essence a group or social phenomenon, like language, religion, or law. Such a view criticizes psychological EP's for casting the individual's relations to others as a balance of internal forces within him; an attempt is made instead to account for the very rise and properties of a self in interpersonal and social terms. Historical emphasis on change similarly is said to be excessive in ignoring those perennial and unavoidable features in all periods that make it possible to treat morality as a persisting structure in human life.

Perhaps the best illustration among twentieth-century ethical theories of a consciously sociocultural type is John Dewey's. It regards man as a natural biological organism possessing basic biological needs; but these needs are held socially patterned. Dewey's ethics is also constantly concerned with psychological questions; but his psychology is itself of a behavioral-social type, rejecting the introspective approach as a residue of dualism. His fundamental concept of *habit* is thus a concept of social, not individual, psychology. It is more akin to custom, as patterned culture in action. In their interaction, habits constitute the self, the will, and character; and all features of the mental life introspectively ascertained are to be explained by them, rather than the reverse. The moral situation itself is defined in terms of "the mutual modification of habits by one another,"[35] and reflective morality arises out of conflicts in habits or customary morality. Purely biological needs can function in the ethical process only as generalized impulse coming to the fore in the conflict of habits. The spotlight in Dewey's ethics comes to rest on the sociocultural.[36] The framework of moral conceptions expresses the permanent contours of associated living.[37] The content of problems, and so of morality, is recognized to be undergoing constant transformation historically, but, instead of moving into a specifically sociohistorical stage-setting, Dewey focuses on the method of successful solution of problems as the perennial clue in ethical understanding.

A considerable part of current sociological treatment of value questions seems to be moving toward a sociocultural EP. The social group is often pictured as engaged in such tasks as maintaining its cultural pattern, adjusting to changing forces, harmonizing its conflicts; sometimes there is reference to more specific but nevertheless perennial problems—continuing its population, transmitting its skills, satisfying its material needs, organizing social relations among age and sex groups, and so on. Ethical processes emerge as control processes and refinement-adjustment mechanisms or cohesion-forces, embodying some degree of deliberation. Such analyses, while carried out in terms of providing social understanding of value structures, are equally available in constructing ethical theories.

Sociohistorical EP's are differentiated from sociocultural not by their recognition of change but by a different estimate of its theoretical significance. They find historical variables or epoch-parameters strategically situated within ethical theory, so that their neglect would constitute a misreading of ethical processes. In a specifically sociohistorical EP, every present social state is seen as a more or less temporary equilibrium of forces in a process of continual change. On this view ethical processes have a directional significance; the particular plot depends on how the historical process is itself read, on what degree of unity it is found to possess, and also on the kind of biological, psychological, and sociocultural subsets that may be associated with the historical setting.

For specifically sociohistorical EP's, Marxian theory in its historical-materialist aspects provides a clear illustration. Here major stress falls on the sociohistorical specificity in the materials of ethics, both content and structure. Given the familiar Marxian picture of historical development in terms of the growth of the forces of production in the history of mankind, the changing systems of relations of production, the conflict of social classes, consciousness in all its forms is taken to reflect the needs, pressures, and tensions of this matrix and the defense, critique, and projection of solutions for the interests involved and emerging. Thus the moralities at any period will reflect—both in the sense of causal origin out of previously existent materials and in the sense of being geared to—the basic needs of the dominant productive modes, the particular form of the relations of production, the particular stage of the struggle of classes (at least during the historical period in which classes constitute crucial social phenomena). While many elements of moral content change with relative rapidity, what appears as more lasting or more perennial or more structural or even as abstract is not free of historical dependence but represents historical recurrences and similarities (though often with changed content) or what is grounded in somewhat longer-lasting features of economic structure or social situations or expresses directional trends in social evolution. To attempt to understand moral phenomena and ethical structures without the full sociohistorical stage-setting is taken to risk distortion. Not only does "Thou shalt not steal" acquire its meaning from the existent property-forms but the idea of a divine command corresponds to a particular stage of historical development, the ideal of justice even in its universal aspects represents the cry against the exploitative atmosphere of class domination, and the hopes and aims underlying it reflect the effort of men for wider productive freedom.[38]

The special force of sociohistorical EP's comes from combination of wide historical sweep with insistence on epoch or period or even particular-moment specificity. Wide sweeps are especially to be found in the integration of biological evolutionary with historical vistas. Thus Herbert Spencer delineates the changes that take place in ethical conceptions as mankind moves through successive stages; or Kropotkin moves from mutual aid among the animal species to a picture of the historical struggles of the co-operative phases of mutual living against the coercive.[39] Julian Huxley's "Evolutionary Ethics"[40] shows many points at which biological, psychological, and historical components can be fused in a broad-vista stage-setting. Thus there is the wider background of the emergence of man with a particular biological equipment; the specific psychological mechanisms by which conduct is charged with feeling and an order maintained through the period of growth; and the historical shift in the very functions of ethics corresponding to the central focus of human problems in the light of the degree of social development—from an ethics of solidarity required for survival, to one of group-domination, to a contemporary task of providing opportunities and safeguarding the possibilities of mankind's future development.

13. Science in Theological and Metaphysical Existential Perspectives

Theological EP's, found in the ethical theories of the supernaturalistic religions, are characterized by the central role given to the concept of the supernatural or the divine. Western religions exhibit considerable differences in their picture of God's properties, man's derivation from God, God's intentions for man. This variety is considerably increased if we go beyond the Western tradition. The ethical potential of God's properties is often considerable. For example, usually God is taken to be more or less on man's side and to have a dependable character. One may conjecture how the ethical process might be conceived if God were construed as omnipotent enemy or as utterly arbitrary in mode of action—the latter a suggestion approached in medieval views of the primacy of will over reason in God. In some primitive religions supernatural action is in fact sometimes seen as arbitrary and even malevolent.

Despite their concern with the supernatural, scientific issues arise in many ways in such EP's, especially in assumptions about the character of human feelings, human responses to sanctions, human cognitive faculties, the causality of human predicaments, the influence of the non-human environment, and so on. The patterning of psychological or social or historical components within theological EP's is often sufficiently marked to differentiate them as theological-psychological, theological-historical, etc. And within each there are marked selective emphases. The score for the orchestration of the human feelings goes from the stress on calculated benefits of eternal happiness as reward for obedience, to filial emotions, and beyond that to refined feelings of respect, dependence, sense of finitude, guilt, anguish. Some recent religious stage-settings are almost wholly psychological in their content. On the other hand, traditional theological ethics has often used a wide historical stage, with God as author of the plot. The story varies from the Augustinian development from creation to resurrection to the belief of a Persian sect that the world is waiting for seven great goblets to be filled with human tears. Variety is also found in the sanctions imbedded in the modes of divine action in relation to man— from emphasis on reward and punishment here or hereafter to theories of grace and the ways in which it may occur. Problems of the extent of control are focused in the range of beliefs about free will and predestination. Questions of the possibility of man's knowing the divine take shape in rational, mystical, au-

thoritarian-revealed and individual-intuitive religious episte-
mologies, with different imbedded conceptions of the character
of the human condition. Theories of environmental influence
also play their parts. Just as Augustine found it necessary to
reckon with the belief that men's fate is written in the stars, so
subsequent religious ethics had to shed the Augustinian picture
of demons and come to deal with psychological forces in human
guilt-phenomena. Similarly today, primitive religions moving
into the modern world have to reckon with the impact of the
knowledge of germs on their beliefs that illness had a moral
punitive character. One of the chief analytic tasks in examining
theological EP's is to sift out what are the scientific questions;
for example, how much of the concept of sin as an experience of
the sinner is covered by psychoanalytic accounts of guilt-feeling
and its configurations. There are no doubt differences of orientation here between religious and secularist approaches to such
questions, but the inquiry of the extent to which science pene-
trates can be common to both. It is worth noting that even questions of argumentation in theological accounts can be considered from this point of view; for example, how far in the
variety of arguments for the existence of God is there reliance
on specific accounts of the nature of the human intellect, of
memory, of the character of knowledge (for example, of the
absolute truth of mathematics), of the moral feelings, and so on.

What is the influence within a theological ethical theory of
the specific answers given to all these scientific questions in its
EP? Research has not been organized along this line of inquiry
to any considerable extent. But the historical comparative picture of religious differences on ethical questions prompts the
suggestion that the influence is a great one. The reach of science, however, may go even beyond this to deal in an indirect
way with other parts of the theological EP. Thus a question like
the existence and properties of God may be considered in terms
of the scientific study of the ways in which these ideas function
in the ethical process in the lives and thoughts of men. For
example, a comparative study of theological EP's suggests that
it is not the mere assumption of the existence of the divine but

the specific properties of the divine that carry the burden of ethical functioning. The Kantian thesis that "it was the moral ideas that gave rise to that concept of the Divine Being which we now hold to be correct,"[41] taken out of its religious argumentation context and viewed as a historical hypothesis, points to the further thesis that the properties assigned to the divine being at any period have in fact been guided by concern with the type of ethical process which will be the outcome. This is a scientific-historical thesis of major importance in dealing with the relations of religion and ethics, but in any case its logical status is that of a purely scientific generalization.[42]

In similar fashion, it can be seen that *metaphysical* EP's do not go wholly beyond the reach of science because they are metaphysical—and irrespective of the nature of the metaphysics asserted. Let us take as an illustration the teleological types. Here the metaphysical elaboration predominates, although a religious element remains at the periphery. The world is set as exhibiting purpose in its structure, and man is given a place in the purposive scheme. He is pictured as fundamentally trying to go in a certain direction, whether conscious of it or not. The Aristotelian stage-setting puts man at the top in a plurality of natures seeking expression. It has biological, psychological, and social components, but is definitely non-historical. The Hegelian stage, by contrast, provides an organized historical progression according to a fixed logic. What the individual is, how he is composed, and what he strives for are referred to the wider pattern. Ethical processes are given a meaning and reference in terms of this framework. While the teleological outlook may be readily incorporated in a religious perspective, with deity prescribing the basic purposes—and was so incorporated in the fusion of the Hebraic-Christian and the Greek philosophical traditions—it must also be remembered that, in ancient philosophy, teleology was also the dominant form that science took, so that it constituted a stage in specific scientific endeavors.

Teleological EP's operate primarily with the notion of a *plan in things*. The plan has a locus either in every individual who has his own nature, or more often in the species every member of which is endowed with the same nature, or in the group or historically continuous people conceived as a single entity, or in the whole of the world as a unitary being. A mode of operation is also specified for the plan; theological teleologies often use sheer fiat or act of will, Aristotelian teleology has nature work like the artist or craftsman with the plan as a kind of immanent blueprint, post-Cartesian teleologies model the plan operation on their theories of the relation of mind to matter, and in idealist (Hegelian-type) teleologies the plan becomes the "logic" or spirit working its way out in the appearances of the world or its historical unfolding. The relation of the individual to the plan takes a different form in each of these types.

The core of the teleological EP lies in the way it employs the human-nature concept. Its concept of nature combines three strands: how an entity acts universally or for the most part (regular or lawlike behavior), what is native or primitive in its constitution (its unlearned behavior or the path of maturation and action in the absence of "distorting forces"), and what it is in some basic sense striving to attain and so (with an implicit definition of 'good' as goal in basic striving) what is the good for it. As long as we assume, say, in the case of man, a stable system operating according to plan, the accounts of regular mode of expression, inherent drives, and basically satisfying activity will yield correlated results. Temporary divergences or irregularities will be seen as accidents rather than mutations, as temporary lags rather than basic changes of direction.

Even the pursuit of evil will be seen as but a distorted way of pursuing the good. Men follow the apparent good, says Aristotle, which may not coincide with their real good; even the devil, says Aquinas, was not naturally wicked, and the will can tend to something only as a good to its nature; the next stage coming into historical being, according to Hegel, must be higher, because it is a next step in the movement of the World-Spirit toward the fulfilment of freedom. In all these forms, clearly, there is assumption of stability in the underlying plan. It is the feeling that this stability needs some grounding which no doubt accounts for the reaching-out of a formalized teleological stage-setting at its borders, either to a religious framework or to a scientific account of basic forces guaranteeing the system as a whole.

Now, whatever metaphysical form the teleological account takes, its picture of stability, of dependable human motivations, of the frustrations consequent on distorting forces, embodies numerous questions requiring scientific answers. (In this respect there is a clear isomorphism between, for example, the ancient teleologies and the contemporary concern with teleological mechanisms, self-stabilizing or homeostatic systems.) And with respect to the elements of a non-scientific type, the sociology of knowledge as a science can study their valuational functioning in the same way as was indicated above for theological stage-settings.

14. Science in Transcendence Existential Perspectives

Our concern here is with theories that disparage the role of existential assumptions in the internal workings of ethical theory and, in varying degrees, propound the autonomy of ethics. For example, G. E. Moore, in his early review of Brentano's *The Origin of the Knowledge of Right and Wrong*, says: "The great merit of this view over all except Sidgwick's is its recognition that all truths of the form 'This is good in itself' are logically independent of any truth about what exists. No ethical proposition of this form is such that, if a certain thing exists, it is true, whereas, if that thing does not exist, it is false. All such ethical truths are true, whatever the nature of the world may be."[43] In his *Principia ethica* Moore fashions the now-familiar concept of the naturalistic fallacy to brand any interpretation of 'good' in terms that are natural, theological, metaphysical, or in any way descriptive of existential entities.[44]

Such a purist approach is not, however, the property of analytic schools alone. In a phenomenological vein, N. Hartmann argues for a realm of value over and above the sensory scientific and theological-metaphysical domains and constantly stresses the purity of this transcending domain, accessible to an intuitive sensibility. In some forms of contemporary existentialism the focus is not on an object of transcendence so much as an act of transcendence, in which the self finds its absolute freedom in the process of deciding. By grouping such varied views (and others) together as a class of transcendence EP's, we imply that there is somehow here a common view of existence and that it furnishes a special type of stage-setting for ethics. But we have the task of showing that in such views the stage-setting is inarticulate or incomplete or even displaced and that, when uncovered, it is found to pose scientific questions.

In Kant, from whom stem most modern claims for the autonomy of morals—that morality is not a function of any existential situation but is unique or *sui generis;* that ethical processes somehow transcend existence and tell it what it ought to be—there is little difficulty in discovering an incomplete stage-setting. The autonomy stress is directed primarily against hedonistic views or any that make obligation a function of desires, passions, sentiments. But there is one feeling Kant wishes to maintain as ethically relevant; this is awe, or respect, which he denies to be a natural sentiment. He is perfectly ready to set broad existence conditions for morality, both in the portrayal of man as in tension between two worlds and in construing his account of the categorical imperative as an exhibition of man's rational nature. (In fact, the account is said to hold for all rational beings other than man if there be such in the universe.) The outcome is simply that Kant is setting a stage but is precluded by his own theory of knowledge from investigating it scientifically. Hence it remains incompletely presented, and there remain large gaps and obscurities in his ethics.[45]

As long as the classical ontological schemes were dominant, transcendence was teleologically tinged, and emphasis was on the object of transcendence to which the individual was rising, whether God or Truth or Reality or the Good. When Kant appointed himself receiver in the alleged bankruptcy of metaphysics, the object moved into the background to be veiled in obscurity, leaving the act of transcendence alone on the stage. It took many forms, beginning with the transcendental ego of Kant's theory. In some philosophers this was the opening wedge for a modern idealism. Sometimes the central focus on transcendence is quite explicit and deliberate. For example, T. H. Green, who stood in the full shadow of the evolutionary theory and its spreading applications in ethics, looked to the epistemological transcendence of man explaining his world, and the moral transcendence in the consciousness of a moral ideal, to stem the tide of naturalism.[46]

The pure-axiological ethics, whether cast in the phenomenological vein or in the British analytic vein, restores an object-orientation freed from traditional teleologies by relying on a cognitive transcendence act. The clue for finding the existential perspective in such ethical theories is to suspect a displacement to the mode of cognition.[47] Neither Moore nor Hartmann rejects the reference to the structure of existence in the *application* of ethics. For Moore, what is right is what will *produce* the greatest good, implying a causal ordering of existence. And, for Hartmann, the positive, as contrasted with the ideal ought-to-be, and so the ought-to-do, will depend on the tension between the structure of existence and the ideal domain. But, with respect to the cognition of the good or the ideal ought-to-be, Moore relies on a kind of self-evidence, and Hartmann on a kind of valuational sensitivity in grasping the ideal.[48] There is a spurious simplicity in their claims: the quality of goodness is there before you, it is self-evident, you either see it or do not, and there is nothing to argue about; the value is grasped by those who have the appropriate sensitivity, not by those who are morally blind or pass by on the other side of the street. It is this feature which prompts the search for existential presuppositions not in the further exploration of the object pointed to or the language of realistic Platonesque ideal entities describing the object but in the mode of cognition. The simple clarity of the insistence on non-natural qualities or ideal objects is unavoidably coupled with vagueness or obscurity in describing the mode of cognition. Hence arises the hypothesis—which I believe a full-length analysis of their ethical philosophies would substantiate—that the EP is to be found in some conception of the self and its cognitive-affective activities.

Contemporary existentialist and existentially tinged philosophies have focused most explicitly and most sharply on the act of transcendence. Reinhold Niebuhr's whole theological and ethical outlook rests on the capacity of the human spirit "of standing continually outside itself in terms of infinite regression,"[49] which indicates an essential homelessness of the human spirit; Niebuhr sees this as the basis of human freedom, creativity, and uniqueness. Although Sartre criticizes both the tendency in the later work of Husserl to turn toward a transcendental ego and the idealist-religious strain in other existentialists, he is himself building a concept of absolute freedom and responsibility on the ability of the individual in choice somehow to transcend all determinants and established guides.[50] Whatever our estimate of these pictures is to be, and whatever the outcome sought by their advocates, there can be little doubt that we have here, in the sense analyzed originally, an EP, but narrowed down to a particular view of the self in action—a striving of man beyond himself, a kind of self-extricating process.

Now the study of the self and its features is in principle a scientific question. It is, of course, possible that there are limits to its scope, but these could be discovered only in its study, not as antecedent postulates. Thus, no matter what form the transcendence stage-settings for ethics may take, such questions as the cognitive beholding or the affective sensibility of man, or, in turn, the aloneness of the human spirit, its activity of rising above or its endlessly regressing movement, its creation of a psychic distance or gap from its object of beholding, and no doubt many more types of subtle phenomena are all serious materials for scientific scrutiny both as phenomena and as elements that enter into ethical processes.

15. Evaluation of Existential Perspectives

The purpose of a comparative study of EP's is ultimately to stimulate the construction of a more adequate one for contemporary ethical theory. This involves evaluation, and so some marks of adequacy. Let us specify a set of terms for such a discussion. We shall call a feature that is appealed to in an evaluative inquiry a *reference point*. A class of reference points is a *standpoint*. So, for example, we can ask of an existential perspective whether it is clearly formulated and whether it is consistent; these are reference points, shaping up into or expressing a logical standpoint. A reference point becomes an *evaluative criterion* when it is assigned a positive or negative value in an evaluative reckoning; consistency almost always, and clarity usually, in a reflective enterprise has a positive value. A unified system of criteria we shall call a *standard*. Sometimes the standard is first on the scene and generates criteria; sometimes the criteria are first, fragmentary and scattered, and are unified by a theory of their relation or by some underlying purpose. We shall meet these concepts later in considering evaluative processes. Here we are concerned with suggesting some standards and some standpoints for evaluating EP's.

Logical standards.—These include criteria of conceptual clarity, consistency, and a variety of points of methodological refinement. These on the whole are sufficiently accepted in most contexts to be presented as a standard.

Many concepts playing a central role in the EP's sketched above lack conceptual clarity. For example, biological-psychological EP's that see human life as a struggle for power are notoriously vague in explicating the notion of power itself. The history of hedonism shows the ambiguities of 'pleasure'— whether a feeling, an act of preference, or a psychological surrogate for some pattern of social activity. Similarly, 'self-preservation' often enters the stage-setting as designating a biological tendency to keep alive, and then shifts in the ethical process to a psychological tendency to achieve one's ideals; only if the connections are established for the transformation by showing how the self grows and changes qualitatively in its effort to survive is confusion avoided, but this means enlarging the biological EP to a biopsychological type.

Inconsistency arises from conflicting properties of the entities involved in EP's. Hedonism has often been criticized for setting the human quest as maximizing pleasures; for, it is said, feelings cannot be summated. The problem of evil in traditional theological EP's in effect charges inconsistency in the description of God as omnipotent, omniscient, and good while allowing evil to exist.

Methodological refinement refers to the way in which a theory is constructed for use—how precise are its operative tests, how clear the relation of its parts, or the distinctions between what are data and what is interpretation, and so on. For example, the concept of human nature in the teleological EP's is often criticized on the ground that it has no dependable tests for the natural when its combined indices fall apart. It was realized after Darwin that the human-nature concept in its account of regularities is systematic-descriptive, not causal-explanatory; that the idea of what is native covers a number of different concepts—now represented by such ideas as instincts and drives, and still requiring careful analysis; that there is no a priori guaranty that any of these "natures" will remain fixed or that they may not contain incongruent tendencies or forces working against one another; that the idea of good may be linked in different ways to man's nature so that whatever is natural is not necessarily *ipso facto* good—in fact, it may be a residue of previous evolutionary developments no longer serving a constructive role.

In similar fashion, EP's that severely limit the span of human sentiments—as Hobbes's view of self-love as covering the generous feelings—often lack methodological refinement, failing to analyze adequately the logic by which other alleged sentiments are to be "reduced" to those admitted on the stage. Or theological EP's that speak in terms of God's will may fail to provide a means of determinately applying the concept.

Failure to distinguish data from interpretation characterizes especially many transcendence EP's that go from exhibiting the transcendence act on the part of the self to some ontological assertion about man and his world. But comparative study shows that the same phenomenon may be variously interpreted: it leads to an "environment of eternity" in Niebuhr's theology; is imbedded by Santayana in a materialist view of man as "a portion of the natural flux" with a moving center and maintain-

35

ing equilibrium by striving for the ideal; or issues simply in a common-sense explication of "the systematic elusiveness of 'I,' " as Ryle carries it out.[51]

The logical standpoint is probably never wholly denied a place among criteria; and, of its elements, consistency is most likely to be admitted, for its absence is frustrating. The degree of clarity achievable and of methodological precision is sometimes taken to rest on the extent of human powers and the orderliness of the subject matter; if so, their admission as criteria becomes a partial consequence of the application of truth-criteria. This indirectly admits them in any case, where the discovery of truth is taken to be a scientific matter, since logical criteria are internal to scientific method. But if the stage is deliberately set to be obscure because the theme and action require candlelight and looming shadows, then it is likely to be associated with a metaphysics of irrationalism, which often, in spite of this, pays its tribute to logic by seeking to exhibit its own reasonableness!

Truth standards.—Since an EP purports to give a picture of ethically relevant aspects of the world, man's nature and condition, the accuracy of any part of the picture may be called to account, and, insofar as science has penetrated any field, to scientific account. Truth is thus imbedded in the aims of an EP; it is not merely seeking to give a pleasing account. (Of course, if the truth of life prove too horrible, some will always be found to shift the valuation and ask for a soothing delusion.)

Differences of degree and shading in truth-criticism are not without significance. An EP may be charged with outright falsity on the claim that its entities do not exist, as the atheist denies the existence of divine beings central to a religious ethics. Or the charge may be partial falsity or misintepretation of data, as critics of Freud's death-instinct would recognize the importance of self-hatred phenomena but deny their underived instinctive basis. Or the charge may be that of taking literally what is only to be seen as mythical or figurative representation— as stage-settings in terms of a "spirit of a nation" determining the focal duties of a people may be said to turn a symbolic expression for cultural tendencies or patterning into a collective ghost. Sometimes the charge combines truth-criteria with other standpoints. For example, the charge may be partial truth, misleading because of what is neglected; this includes also a failure in comprehensiveness. Or the indictment of "ideology" charges some falsity combined with the subserving of narrow or personal interests, sometimes with the element of self-deceit. (Here we sometimes find the language of criticism which speaks of a "false consciousness" interposed between the observer and the real phenomena, or the psychological language of a "screen," or the sociological analysis of a "lag'" in intellectual habits.)

Of special interest is a particular truth-criticism which we may label as "lacking independence." This characterizes an EP whose sole evidence lies in the phenomena of the moral domain. Thus Kant sets the stage with man as having free will; but, according to Kant, the only evidence for it is the demand for it implicit in the moral consciousness. Similarly, if an EP included the picture of man as inherently altruistic in a world of completely selfish action, and offered as evidence that the morality embodied the injunction to love thy neighbor and was meaningless without a deep-seated human-nature base for it, it would lack independence. This does not imply that moral phenomena do not constitute evidence, and one could conjecture about what a world would be like in which they would be the sole evidence. But, clearly, an EP that is independently established is in a much stronger position. Psychological and theological EP's have both been charged at various times with lacking independence, but their response has tended to be different. When it is claimed that the pictures of man as selfish or aggressive or cooperative and naturally sympathetic have been geared to the advocacy of particular social outlooks in the history of political theory, psychology as a science attempts to shed these elements as value-intrusions and insists on clearer empirical marks of the traits involved and, where possible, experiments to determine properties of man's nature. In religious theory, on the other hand, the growth of self-consciousness about the gearing of religious EP's to ethical demands (the hypothesis indicated above) has often been taken to point to the center of the religious outlook.

The standpoint of comprehensiveness or completeness.—One EP may be more complete or comprehensive than another. This is not so simple a standpoint as it seems, for there is always the question: comprehensive enough to accomplish what? We could answer by referring to "the tasks of ethics" and so speak of a standard of comprehensiveness; but, since the tasks have not been systematized in contemporary theory, it is more prudent to limit ourselves to speaking of a standpoint.

In general, the situation is analogous to that in a play: there is no virtue in sheer crowding of the stage, but the stage is not comprehensively set if the action requires props which are not there. There is a clear sense in which—granted they are both true—a stage-setting which describes cultural properties in addition to psychological properties is more comprehensively set for application to moral processes. On the other hand, an EP that is less comprehensive—for example, a physical or biological one —may maintain itself by a promissory note, that the qualities of human life and personal relations which are the common currency of ethics can be furnished by future correlation to physical and biological processes. This would seem to show that comprehensiveness is a function not of the number of entities involved in the stage-setting but of their properties and theoretical explanatory power.

Since comprehensiveness as a criterion is relative to some assumption of the jobs that ethics has to perform, a circle may arise which it is well to note explicitly. The conception of the tasks of ethics is itself part of a fully expanded ethical theory and so is probably to some extent a function of an implicit EP. Hence the point may be reached where a narrow EP justifies its own incompleteness by narrowing the tasks of ethics. A good example is to be found in G. E. Moore's ethics. In his denial of the relevance of the character of existence to ultimate ethical judgment he has in fact made the stage so bare that the only kind of action possible is ethical star-gazing, not ethical navigation. The tasks of ethics are narrowed to knowledge, not practice; few procedures are developed for transforming acts of vision into methods of guiding human action; and it is then found on the basis of the constructions offered that ethics is practically impotent to challenge any moral rule that happens to exist.[52] Another example on a large scale is the modern narrowing of ethics to personal rather than social issues. Its individualistic stages— usually of a psychological type—so narrow the conceived tasks of ethics that little relevant advice can be given to social policy.

Fortunately, from a theoretical point of view, the circle is rarely completely closed. The pressures that generate moralities are insistent enough to keep a wider notion of the tasks of ethics open, so that comprehensive adequacy remains a workable criterion. The comparative conclusion that EP's constructed in narrow terms tend to broaden out to cover a wider field— whether by shifting the meaning of their central concepts or by spreading wider their theoretical nets (as, for instance, we saw, Freudian theory moved out over social relations and even historical development)—and the historical lesson that theories offered first in a limited way tend to pick up supplementation (as Utilitarian ethics in the 19th century gathered a biological base and a historical scope in the evolutionary utilitarians) both tend to suggest that comprehensiveness can be regarded as a criterion implicitly admitted even where not explicitly acknowledged. But even in the extreme case where the circle described above has been closed, the issue is simply shifted directly to the varying conceptions of the tasks of ethics—a partly historical, partly psychological, partly policy-decision question.

Orientational and functional standpoints.—There are many interesting and potentially valuable approaches to EP's stemming from psychological, social, and historical study of their impact and modes of functioning. These are perhaps unified enough to be regarded as constituting an orientational and a functional standpoint.

The orientational standpoint asks questions that are perhaps less causal than phenomenological. How, if men set the stage in a given way, would they tend to feel and act in making ethical judgments? Would the EP carry with it an active outlook, or passivity and resignation, or a constant feeling of being shoved around? Would it prompt a rational process or a mustering of feeling and pressure? Do religious EP's prompt to resignation in the sense of consolation or of acquiescence? Do they have an authoritarian potential or a liberating quality? On a teleological EP a man may have the exalted sense of being drawn to the light. On a mechanistic one he may have a sense of being propelled, whether the forces be the impact of molecules, neuromotor reactions, conditioned reflexes, repressed desires, or propaganda. Karen Horney criticizes Freud's theory as leading analyzed people "not to take a stand toward anything without making the reservation that probably their judgment is merely an expression of unconscious preferences or dislikes" and as jeopardizing the spontaneity and depth of emotional experience.[53] The more comprehensive naturalist EP's convey the sense of a man in continual dynamic interaction with his surroundings, physical and social, and regarding his choices as a responsible fashioning of himself. In general, then, the orientational standpoint involves generalizations about attitude-tendency. It thus specializes in what we may call the *virtue-potential* of a stage-setting.

Such judgments are by no means simple. There are rarely one-to-one relations between the stage-setting and the attitude associated with it; fatalism may bring passive acceptance or a destiny-activism. The actual virtues that emerge in the ethical theory are a function of the whole picture and not of the stage-setting alone. For example, an EP whose general orientation is to realistic appraisal of existing conditions may yield stoical acquiescence in a world that is realistically discoverable as hard but quite different virtues in a world that is realistically found full of opportunity.

The virtue-potential of an EP may be explored with respect to each of the EP dimensions mapped above, such as the space-time features or the steadiness and intelligibility of the world, or it may be related to configurational properties, such as the qualities of being well equipped or at a loss, or the degree of room for action. William James treats a free-will doctrine in such a fashion when he identifies it as a general cosmological theory of *promise*.[54] Similarly, his attacks on the notion of the Absolute and his advocacy of a pluralistic approach were directed primarily against an attitude of a closed world to which one could only become resigned, in favor of an open world with room for human initiative.

There is no sharp break between the orientational and the functional standpoint of inquiry, but there is a marked difference in emphasis and direction. While the orientational standpoint is largely phenomenological, the functional is concerned with psychological, social, and historical roles and services and is more intimately related to causal-explanatory inquiries. Thus, for example, the question raised whether a specific theological EP involved acquiescence or consolation was orientational; if we probe into psychological relations and ask whether it releases guilt or intensifies it, whether it functions as a projective system or as a realistic facing of the totality of things, or into the history of religions and ask whether they serve as specific opiates or vigorous social organizers, we are following the functional path. Similarly, it is an orientational question whether a physical EP carries an atmosphere of promise or of coercion; but it is a functional inquiry, resting on causal analysis, to differentiate the optimistic eighteenth-century mechanism, with its promise of a clean slate and the remaking of man, from the twentieth-century pessimistic counterpart which identifies being a machine with being manipulated.

Inquiry along such functional standpoints is clearly scientific in type and method. All the relations between belief and feeling that a developed psychology may discover and all the lessons of the theory of personality and its mechanisms may have application in studying the psychological functioning of EP's. Comparably in sociohistorical functional inquiry we have to ask how the stage-setting fits into group aims, whose aims, and distinguish ascertainable real aims from apparent aims. Thus a hierarchical picture of the cosmos as a basis for ethics may in a particular age and society serve the social function of a fixed and stratified feudal system. Similarly, an equalitarian hedonistic stage-setting, with every man pictured as knowing best what will give him pleasure, may function socially to support and justify a laissez faire economy. Only a scientific functional analysis could explain why the teleological idea of the natural carries a conservative acquiescence pattern in one age and the idea of the native or inherent a revolutionary pattern at another time. Similarly, analysis would have to distinguish where the social conditions operated as cause of the EP's adoption, where a pre-existent EP was put to a fresh social use, and where the social function was itself a constitutive part of the EP.[55]

Whether such orientational and functional standpoints can generate evaluative criteria and standards is a distinct issue. In certain contexts they could quite readily; for example, where a given end was widely accepted, they could provide criteria for the *successful functioning* of the individual and the social field with respect to that end. In general, they can furnish standards only to the extent that there develop evaluative concepts of *a healthy personality* and at least minimum agreed-on concepts of *social well-being*. But these in turn presuppose well-established scientific theory, as well as some shared human purposes.

If the evaluation of EP's takes the indicated shape, the place of scientific inquiry and scientific results in the fashioning of an EP for contemporary ethics is a large one. The EP raises precisely the kind of questions for which scientific answers are required, even if the roster of the sciences of the day cannot yet answer them. How far an EP will also require some components of a non-scientific type (metaphysical, theological, etc.) is likely to remain the subject of controversy. The more sanguine may anticipate a situation in ethics comparable to that in biology, where non-scientific formulations were forced back outside the field into the status of philosophical speculations rather than active participants in internal decisions. The less sanguine may expect that limits to the reach of science in ethics will be discovered by science in its own forward movement. At a minimum, it is likely that the area of scientific answers will become increasingly central within the operations of ethical theory. For example, it will not be a metaphysical or theological interpretation of the human will, of emotions, or of self, that will bear the theoretical burden in ethics but the properties of such processes discovered in psychology and the social sciences, forcing their recognition on all alternative interpretations.

As to particular sciences, it looks as though most of the purely physical and biological EP's that have been offered as self-sufficient for ethics are inadequate in terms of the criterion of comprehensiveness. Even contemporary biologists have stopped talking of biological ethics in nineteenth-century fashion and think more in terms of biological bases for ethics. The psychological components in an EP are of central importance, but, whatever form they take, it does not look as though they can be cast purely in introspective terms or purely in behavioral terms or purely in phenomenological terms. The psychological concepts entering into EP's will probably have to represent configurations of various of these elements (including depth elements) resting on empirical correlations and theoretical unifications. The issue between a psychological EP and a social EP (whether the latter be sociocultural or sociohistorical) is not as yet conclusively resolved. If it is individualistic-psychological, it must embody a clear theory of interpersonal and social relations; and, if it is social, it must make unambiguous room for individual mutation and variation. Favoring the former is the importance of individual choice and decision as a central phenomenon in the field of ethics; favoring the latter is the fact that morality, like law or religion or science, is a social form.[56] But the underlying scientific issue is whether an individualistic psychology is really possible, or whether the inroads of social and cultural materials in the explanations of individual behavior and development do not call eventually for a more integrated model of man overcoming the present dichotomies. From this point of view, the formulation of a contemporary EP cast in scientific terms need not bind itself to the present mold of the sciences of man; it can follow in outline the direction in which it sees the sciences to be moving, although it may thus leave gaps in its detail and so perhaps be compelled to leave some theoretical issues in ethics with sets of alternative answers.

Again, it looks as though any form of EP will have to reckon with the fact of change in human life and, in dealing with every aspect—whether discourse and usage, feeling and motivation, institutions and cultural forms—be attentive to the possible impact in both the content and the structure of ethics of directional transformations. This means that eternities, universals, absolutes, structures, if they are to be offered in the EP, will have to take the form of invariants, grounded or evidenced constancies, methodological necessities, or even stipulated values.

Whether, if a social EP is decided on, it will have to be specifically sociohistorical or can remain generally sociocultural is also an undecided question. Probably it depends on the rate of change to be found in the human field. If there are structures relevant for ethics that have remained constant over the whole of human history and are likely to remain so, and, if they are central enough, then a sociocultural EP may be soundest policy. If there have been fundamental modifications in basic structure, then a sociohistorical EP is better policy. In any case, a sociohistorical EP would have to be formulated so as to allow for possible generalization across societies and epochs; and a sociocultural EP would have to allow for variations in specific detail and fashion subsidiary categories for determining relevance in time and place. Such questions of the extent of invariance and its grounds, although they point to different methodological policies in the present structuring of the field, do not constitute issues of principle. They are determinable questions of science and history, even though the answers may be long in coming into sight.

Since the concept of an EP as a systematic tool of analysis in ethical theory is only being fashioned, theses about such outcomes of evaluation cannot here be carried further. But, in the light of even the present state of knowledge, we cannot be satisfied with anything less than a unified EP which integrates the biological and psychological with the sociocultural and historical. The chief obstacle is the present lack of integration in the sciences themselves. This is, however, a scientific problem of which they are clearly conscious.

III. The Role of Science in Conceptual and Methodological Analysis

16. Conceptual-Methodological Frameworks

In contemporary ethical theory we often find questions of the following sort:

What are the distinctive terms to be employed in moral discourse (e.g., 'good,' 'right,' 'wrong,' 'ought,' 'virtue,' etc.)? How are they to be related and moral sentences formed out of them? (For example: "Is 'right' to be defined in terms of 'good' or are these two independent terms? If 'good' is used as a predicate term, is the subject-form unrestricted, or can only an experience properly be described as good? Are 'ought'-sentences really disguised imperatives?)

Can we properly define moral terms by non-moral terms? What kinds of relations are to be permitted between moral terms and descriptive terms referring to experiences, feelings, phenomenal qualities, contexts of human processes, and so forth? Are we to allow equivalence-definitions or other types of semantical rules? Or is the relation to take the form of some specified contextual function (e.g., to express feeling or to commend)?

What relations of moral sentences are to be permitted? Are there logical relations such as consistency between moral utterances or more generalized material relations of coherence? What form does organization or systematization take within a morality? Are there laws and systems of laws? Or, in some other sense, hierarchies of norms? Or rough collections of discrete decisions with at most family resemblances? Or other forms of patterning?

What modes of certification do moral sentences allow? Is it some type of cognition or empirical verification, or some type of feeling-sensitivity, or some type of willing or commitment-acceptance?

What modes of reasoning and justification are appropriate to the moral field? Is ethical reasoning deductive in form, or inductive? Or has it its own type of logic, with its own criteria of 'good reasons'? Is there a logic of choice as distinct from a logic of thought? What is the nature of the process of application and decision? Can decision be rational or do decisions ultimately "just happen"?

Answers to these and hosts of kindred questions about how the concepts and methods of morality are to be analyzed, construed, employed, and applied give us a fairly clear indication of the *conceptual-methodological framework* of an ethical theory. They furnish a kind of *logical profile* of that theory. Comparative study of logical profiles raises questions of their evaluation, and attempts to reconstruct the methodological framework of ethics call for framework policy decisions.

Our inquiry here is twofold: (*a*) How far are specific results of the sciences presupposed in the occurrence or adoption of a specific logical profile in an ethical theory? (*b*) How far is it possible to employ scientific method in ethics, as against regarding it as an overextension of a misleading model?

a) Scientific results were seen to play a large part in the constitution and evaluation of EP's; such influence carries over into frameworks if logical profiles take different shape according to the specific EP imbedded in the ethical theory. In the preceding chapter we started with EP's and suggested the scope of their influence; in the present chapter we start with frameworks, utilizing illustrations of framework problems, and look for points of indispensable EP reference. For example, one traditional kind of theory links the meaning of 'good' to *self-realization,* another to an act of *commitment,* another to *pleasure.* These may all be regarded as *EP variables,* since their meaning is furnished by the specific account of the self, of the will, of pleasure. (The picture of the self in self-realization theories of modern idealism differs markedly from that in contemporary psychoanalytically grounded ethics; even 'pleasure' in the hedonistic tradition has different interpretations.) The heuristic principle in the inquiries of this chapter is that whatever framework problem is analyzed we will find its answer to be in some significant measure grounded in the values assigned to the EP variables in the ethical theory. There is no one way in which scientific results in an EP enter into all frameworks; it is the task of comparative research to sketch the actual variety with which scientific approaches and judgments embodied in different EP's enter different logical frameworks. Nor is it being maintained that scientific considerations are the sole determinants of the logical profile. Comparable study would have to be directed to language-habits and to pragmatic or purposive elements.[57] But it is clear that they enter unavoidably with considerable determinative force and that they have accordingly an important place as grounds of policy decision in framework reconstruction.

b) Under what conditions would it be possible to structure ethics as a scientific enterprise, rather than as an art-enterprise, or a practical enterprise of some sort? Under what conditions would this not be possible?

For scientific method to be applicable in ethics, the human field must be sufficiently determinate to provide the following: (i) A concept of *moral phenomena* (including *moral experience*) sufficient to mark off an area of inquiry (observed qualities, feelings, determinate will-acts or behavioral processes or human interrelations in definite types of contexts) either as distinctively moral or at least as an area in which interpretation of moral terms is to be sought or pointer-readings for verification of moral statements to be discriminated. (ii) A set of moral terms and definite ways for linking them to the established area of moral phenomena. (iii) Some meaning for generalization or systematization in the reiteration of experience or phases of experience or some more complex invariance or descriptive patterning. (This is the condition of possible regularity or "law-likeness" resting on some isolability among the phenomena.) (iv) Some mode of verification or certification for the generalizations and some procedures of validation or justification for working principles of a higher order. (v) Some modes of application and decision so that the systems of generalization wi." have relevance to the practical tasks of morality.

Even if moral utterances were wholly expressive or "blind-volitional," scientific method in ethics would be possible if there is determinateness and lawfulness in the occurrence of the expressive utterances or blind-volitional acts. If a high degree of such regularity were found and a systematic explanatory theory developed, a concomitant or correlated descriptive use of moral terms could arise, just as constructs together with operations replace initial quality terms in any of the customary sciences. On the current philosophical scene, ethical theories of an emotive and prescriptivist type have tended to grow conceptions of validation or of "good reasons" even while insisting on the practical nature of ethics and the practical interpretation of ethical concepts.[58] This amounts to recognizing that there is a degree of determinacy in the field; where it is overlooked in one part of a theory it will come up in another part. In general, if a conception of the task of ethics as practical is offered, the question whether the practical task can be most effectively carried out by a descriptive or theoretical or practical interpretation of moral terms is not itself a *practical* but a *scientific* question guiding framework policy.[59]

The denial of the possible utility of scientific method has to establish the kind of conditions in the field—for example, an intrinsic arbitrariness in the will—which will rule out any way of satisfying the conditions stated above. Dostoevski thus points to "an interest which introduces general confusion into everything"[60] and speaks of the independence of the will at all costs, which may even mean a man acting against his own interests! To determine whether such an interest exists and probes deeply into the nature of man or whether it is a clinical symptom would appear to necessitate a scientific scrutiny of its bases of operation.

The five sections that follow deal with each in turn of the conditions for the possible use of scientific method in ethics. At the same time, however, the discussion of framework questions is oriented to discovering the pivotal role of EP variables in framework decisions.

17. Are There Workable Concepts of Moral Phenomena and Moral Experience?

Consider, for example, the following familiar utterances:

It was my responsibility. My conscience won't let me do it. It is obviously the thing to do. It was a strong temptation, but I resisted it. It was a strong temptation, so I succumbed to it. I felt that was no excuse. I recognize he has a claim on me but. . . . Here I take my stand; I cannot do otherwise. That's outrageous. What else could a man do and still live with himself? That's unfair. I sympathize with his predicament. Have you no scruples about doing it? Have you no compunctions? That would be giving up what I've worked for all my life. It was a courageous thing to do. You'll never regret it. That would be a wonderfully satisfying way to live. I was so ashamed of myself. I felt as if I had been dragged through the mud. Surely we are deeply committed to this. Would you want your child to be like that? It's not worthwhile doing. That was a fine experience. I can see that it is his ideal, but it does not attract me. It's a matter of simple loyalty. I did promise, so I shall do it. After all, I am a member of the group, so I shall bear my share of the costs.

Such a list could be continued indefinitely, moving off in different directions. There are hosts of simple valuings—enjoyments, delights, and satisfactions and their opposites. There are diverse sets of feelings—varieties of guilt, shame, awe, respect, indignation, gratitude, sympathy, care. There are classes of interpersonal reactions—admirations and recriminations, with hosts of finely shaded adjectives. There are apprehended qualities of experience, such as finding something congruent or fitting, or frustrating, or ominous and overshadowing. There are moral-model relationships, such as the experience one has in regarding someone as an authority or as an ideal-figure. There are reflective experiences, such as what one would have chosen if he had had a clearer view or been less excited or what he would have recommended if he had been more disinterested. In all these cases we could ask what kind of experience is taking place—what kind of tasting or perceiving or feeling or willing or reflecting. Or we could ask what kind of phenomenon is taking place—what kind of qualities are appearing, what object-relations existing, what personal relations being manifested, and so on.

How sharp is the demarcation and the articulation of this realm? There have been many attempts to specify a single mark of the moral, as if moral experiences when isolated would be as simple as the sound of a bell or as unique as the taste of a persimmon. Underlying EP's sometimes have a limiting effect on the kind of data explored. Thus the various individualistic psychological EP's in dealing with obligation experiences tend to look inside the individual, and come out with different types of guilt or shame or remorse feelings. Interpersonal and social EP's identify obligation experiences as directly transindividual and so concentrate on claims and counterclaims, rights and corresponding duty relationships, and recognition of whole-institutional demands. A comparative perspective is required to insure extracting the full range of phenomena. Moral experience might turn out to be a complex orchestral experience with many instruments playing and even with cultural variations in the score.[61]

The field of relevant phenomena falls into fairly distinct groups: (*a*) *the desire-aspiration-satisfaction* group, including acts of desire, striving, goal-seeking, aspiration, pleasures and satisfactions, and so forth (and their opposites), as well as the recognition of the objects of these acts; (*b*) *binding-authority* phenomena, including consciousness of demands and claims and ties, as well as feelings of remorse and promptings of conscience; (*c*) acts of *appreciation and depreciation,* including reflective reaction to persons and traits, acts and situations.

These groups provide sufficient basis for interpreting moral terms and for furnishing pointer-readings in verification. This does not have to wait for a fully developed account of precisely what falls within the moral domain. A carefully identified phenomenon—an act of approval, a feeling of guilt, a feeling of satisfaction under controlled conditions—can serve as a verifying instance for a particular moral statement, whether or not it ranks as a moral experience, just as a pointer-displacement verifies the presence of an electrical property without being itself an electrical event. However unsettled the precise marks of a moral experience, gathering a wide pool of possibly relevant experiences is a firm starting-point for extending scientific method in ethics.

18. Ethical Concept-Families and Their Existential Linkage

It has long been recognized that there are three major families of concepts in the ethical linguistic community. One, including 'good,' 'bad,' 'desirable,' and the like, may be called the *good-family*, although nowadays it is perhaps more common to speak of value-terms. The second, including 'right,' 'wrong,' 'ought,' 'duty,' etc., may be called the *obligation-family*. The third is the *virtue-family*, with its broods of specific virtues and vices.[62]

Some existential linkage for such ethical concepts is a necessary condition for the applicability of scientific method to the ethical domain. This is too often discussed as if it required equating each ethical term such as 'good' with some lower-order descriptive predicate such as 'pleasant' or 'is desired' and as if the rejection of such an equation ended the possibility of scientific method in ethics. This is an undue restriction on inquiry.[63] It is also unduly entwined with controversies over descriptivism and non-cognitivism, that is, whether an ethical statement is a descriptive report or serves some other function. The whole inquiry of the existential linkage of ethical concepts acquires wider scope by attentiveness to variations within the ethical tradition and to comparable problems in the philosophy of science generally.

Historically, different ethical concepts have always been closely associated with the various groups of moral experiences indicated in the previous section: the good-family with the desire-aspiration-satisfaction group, the obligation-family with the binding-authority phenomena, and the virtue-family with at least a large part of the appreciation or reflective-approval group. Hereafter, let us use the term 'domain' to cover a group of phenomena as associated with a family of concepts. We have thus the good-domain, the obligation-domain, and the virtue-domain.

We can also note historical shifts in the dominance of the concept-families. In ancient times ethical theory confidently assumed that, if we knew the human good, everything else would fall into place. Medieval ethics seems to have thrust contractual ties and mutual obligations into a more prominent theoretical position. Kantian and post-Kantian ethics have intrenched the concept of obligation as almost definitory of ethics. Virtue concepts made inroads in ancient times by such devices as the Stoic construction of virtue as the primary content of the good and in some modern periods by the central place given to the moral ideal of character and personality. The historical careers of the concepts appears to reflect the relevance of different moral experience groupings to the institutional and historical problems of the day.

Comparative inventory of analyses offered in various theories for each of the central concepts shows that there is always a reference to some portion of the field of moral phenomena. For example, we find obligation analyzed as a voice of veto or command (Socrates' demon, or typical accounts of conscience as a still small voice)

a sense of "office" or a job to be done (Stoic)

awe or respect for law or rationality (Kant)

a sense of overwhelming pressure (one of Bergson's two senses, assimilating it to habit)

a sense of aspiration or attraction (the other of Bergson's senses, assimilating it to the ideal; Plato's analysis of the tug of the Good)

a sense of a governing whole, or a choice by the whole self (Bosanquet, modern idealistic philosophy)

a type of debt (Nietzsche)

a contractual-type of commitment (Socrates in Plato's *Crito*)

a type of reasoning directed to maximization, or to harmonizing conflicting aims (e.g., Bentham, Santayana)

a type of sentiment, such as a pyramiding of sympathetic responses (e.g., Adam Smith)

a sense of loyalty or commitment (Royce)

an anxiety embodying developmental derivatives (Freud)

a vectorial quality of requiredness (Köhler, Mandelbaum)[64]

and, of course, many other ways. The formal features of each conception seem to reflect the material properties of specific moral experiences, in some cases with rich content, in others with only abstract outline.

Analytic controversies, historical considerations, and comparative lessons combine to suggest that it is worth differentiating more systematically the various ways in which ethical terms may be linked to existential entities, qualities, and procedures. Well-developed accounts of such problems in the philosophy of science prove helpful. We find at least five different types of linkage which throw considerable light on possibilities in ethics.

(i) The equation of an ethical term with a descriptive predicate so that we have the necessary and sufficient conditions for applying the term. This is currently rejected as not feasible, but it must be allowed to remain as a possibility. For example, on a descriptivist approach there might turn out to be a unique phenomenological-field property, such as requiredness, equated precisely with a basic use of 'good' or 'right' as a fundamental term. On a non-cognitivist position—say, an expressive or emotive theory—there might turn out to be a unique feeling or emotion for each different basic ethical term (e.g., horror for 'wrong,' approval for 'good'). Although not formulated logically as a definition, it might be offered as a precise model for analyzing the situation in which the term is properly employed. Lewis Feuer proposes that, "corresponding to different social structures with their different personality-forms, there will likewise be diverse ethical languages each with its specific psychoanalytical characterization."[65] For example, for 'This is good' he suggests: for a liberal society, "I like this, and, since we are so much alike, you probably would like it too"; for a Calvinist society: "I dislike this, but was compelled by my father to accept it; now, having identified myself with him, whatever resentment I harbor against him will be deflected toward my own children, who will suffer as I did." Whatever the adequacy of such an account, the attempt itself shows that the search for definition surrogates for ethical terms is not to be barred by an expressive non-cognitivism in ethical theory.

(ii) Non-ethical terms may be offered as an interpretation or model for a system of relations in which ethical terms have already been elaborated, as a physical model may interpret a set of geometric postulates. Though ethical theories have not explicitly employed such procedures, analysis of the relations of ethical terms is sometimes of this sort. Let us construct a simple hypothetical example, with three fairly familiar stipulations about the mutual relations of selected ethical terms:

1. If a person has a *duty* to perform some act, there is some person or persons who have a *right* to its performance.

2. If a person has a *right* to some act, then there is some *good* which he will derive from its performance.

3. Every *duty* has a *ground* in some character of the particular situation or previous situations.

One model which readily suggests itself is that of *debt*. To have a duty may be interpreted as to owe a debt; to have a right is to be a creditor; a good is some proprietary object or service which is the subject matter of the transaction. The ground is the "value" conferred in the loan. This model appears to satisfy

53

the three "postulates." Whether it proves adequate in the long run depends, of course, on the relation of these three to the rest of the ethical theory—additional stipulations, factual assumptions, etc. And, since these are generally rarely worked out explicitly, there is a large area of arbitrary employment of models.

That an ethics in terms of debt can be worked out in detail is clear from its actual occurrence, for example, in Japan, as described by Ruth Benedict in *The Chrysanthemum and the Sword*.[66] There are infinite debts and finite debts, debts repayable only in kind and those repayable in money, etc. Note, in general, how a debt ethics would have to reconstruct familiar obligations in our society. The first postulate demands that for every obligation we find a creditor; the third, that it involve a present or past value received. Thus social obligations would have to be construed as debts to God, to ancestors, even to society collectively for its role in fashioning the individual. Obligations to the future are ruled out, because future generations have not given us a value received. Presumably they could be construed as loans; actually, in Japanese ethics obligations to one's children are seen as repayment of debts to one's parents.[67]

It is, of course, possible that an alternative model may be offered for the same postulate set. For example, *contract* would also be possible, since every debt itself could be regarded as a contract, with the parties as debtor and creditor, the object intended the good, and motives of the parties the ground. A contract interpretation would probably considerably liberalize the obligation system; historically, in the Western world, it has involved a wider individualism.

(iii) A third linkage of ethical terms to existential materials may be compared to the role of operational definitions. An ethical theory may co-ordinate 'X is wrong' with a first-person introspective operation such as 'When I contemplate myself as having done X, I feel remorse.' This does not give the full meaning of 'wrong' but provides a means of partial identification after which fuller exploration of the material identified can then be carried out.

Some traditional ethical analyses may be construed along these lines. In the good-family the test of what is desired for its own sake is perhaps the most prominent. If this is thought of as a kind of operational definition, then we can see the specification of normal conditions—that one must not be in a disturbed condition, or ignorant of what one is doing, or have built up contrary habits, etc.—as comparable to making sure that measurements of length are reckoned at standard temperature and pressure. In the obligation-family the repeated attempts to refine feelings of conscience and remorse, to differentiate them from fear of consequences or pain of loss or hurt to self-love or self-esteem, may be seen as sharpening "pointer-readings." In the virtue-family there was developed the notion of the "ideal spectator" who has his sympathetic and other reactions in a cool hour or from the vantage-point of a "disinterested" observer.

(iv) A fourth mode of linkage lies in the discovery of *empirical* indices correlated with the application of previously established linkages for ethical terms. Suppose 'wrong' is linked to the (operational) remorse procedure. It might then be found that, in some domains of publicly performed actions, social disapproval was a fairly regular accompaniment of acts which produce remorse. Without confusing remorse with fear of social disapproval, it might still be possible to use social disapproval as an empirical index for 'wrong' in that domain.

An operational specification may be outworn and take its place as an empirical index for limited domains, as when a child acquiring a wider understanding of obligation than simply 'what his parents emphatically demand of him' may continue to use this test as at least a first index of duty. Or an empirical index may prove more reliable once a general theory has been developed and so take the place of the original operation with which it was empirically correlated, as with most reflective persons the feeling of remorse itself becomes more of an empirical index than an identifying operation, and the fact that we have hurt others may move more into the role of operative test.

(v) A fifth type of linkage may be extracted from the work of the linguistic analysts, in spite of the fact that they usually regard it as exhibiting the practical rather than the scientific character of ethics. This type joins ethical terms with functional contexts. Ethical terms are treated as having jobs to do; different ethical expressions may be doing similar jobs or the same ethical terms doing different jobs in different contexts.[68] For example, 'ought' may have the function of criticizing, advising, deciding; it may have different significance where a person is deciding what he is to do ('I ought') and where he is specifying a role ('A man in such-and-such a position ought to . . .'). Ethical terms may vary in contexts of general social legislation and individual operative applications, peer-age groups and cross-generation groups, action-problem contexts and educational policy contexts, as well as spectator contexts and participant contexts; and so on. A realistic job-mapping involves a thorough understanding of what is going on in a field in all its institutional and cultural background. In a sense, this becomes the application of an anthropological orientation to the analysis of specific moral patterns.

The five types of linkage by no means exhaust the possible existential connections of concepts. All sorts of other types may play important parts. For example, in frequent explications of 'duty' as an action that is productive of the greatest good, material concepts of consequences or effects enter into the very relationship of the two ethical terms. And in such a familiar slogan as " 'Ought' implies 'can' " there is the suggestion of necessary material conditions for the application of an ethical term, already canvassed in the conception of a stage-setting. But, whatever further developments are possible, the five types indicated open a wide path for a treatment in ethics quite close in spirit to that of scientific inquiry.

19. Organization, Generalization, Systematization

Because the familiar concept of *moral law* has occupied a central position in modern ethics, an effort is required to look at the domains of moral experience and their conceptualizations and ask whether different patterns of organization are possible and appropriate and where lawlike generalization fits or does not fit. Let us examine the possibilities in each of the three major ethical domains.

(i) The good-domain has experimented with several organizations, reflecting the particular stage of scientific study of its phenomena—will, desire, feeling—in short, the complex history of psychology. The *means-ends* category emerges as the general way of systematizing goal-striving. Under classical teleological psychologies it is given a *hierarchical* specialization with the supreme end as the highest good. Obligation phenomena fall into place as means to achieving the good, and virtues represent character-traits similarly oriented. Comparative value is identified with degrees of completeness in achieving the good. Generalizations are possible within this scheme; they are universal or for-the-most-part statements about men's striving for goals or the frequency of success of means in leading to ends. As teleological underpinnings (in fixed human-nature metaphysical EP's or in theological EP's) give way in modern times, this scheme of organization is somewhat transformed. With the removal of fixed ends, the hierarchical character suffers. A number of different types emerge. Some experiment with *part-whole* relations, as in the idealist organic philosophies in which the completeness of the whole replaces the final end. Some develop the concept of *ideals* as organizing and guiding foci. Some turn to the biological aspects of the scientific picture of desire phenomena and fashion concepts of *underlying drives* or *guiding propensities*. A great many model their mode of organization on satisfaction phenomena; hence the great variety of affective theories of value. Of these, pleasure theory is the most prominent, developing its ethical concepts on what it takes to be a universal theory of motivation. Comparative value becomes a scale of measurement of greatest happiness, or maximum preferences. Obligation phenomena are attuned to maximum-generating instrumentalities, and virtues become the safest educational self-investments yielding the steadiest happiness-return. Finally, on the contemporary scene, as new conceptions of the relations of phenomena arise, altered schemes are proposed. Scientific conceptions of homeostatic or self-stabilizing systems have provided a model for combining both goal-seeking and regulative aspects in a more coherent way, although unable as yet to integrate the phenomenal and the affective or to conceptualize shifts in basic goal-direction that are found in human life in historical change. In any case, it is becoming increasingly clear that an adequate organization of this area can follow only on an adequate unified theory of man.[69]

(ii) In the obligation-domain the dominant conception of

moral law traditionally fused authoritarian command, rational order, and universal generality. (In Kant, for example, the content of command turns out to be to universalize, a rational being is identified as one who governs himself by law, and the form in which morality issues is that of laws.) This particular amalgam may represent a modern Western specialization in the pattern of conscience. However, an authoritative command need not be universal in form.[70] Rationality permits of a variety of forms and even involves some appreciation of the conditions of uniqueness. And the very idea of moral laws in the sense of obligation-universals is only one of a variety of forms that obligation phenomena and their analysis may bring to light.

If laws be studied as types of rules, then at least the following may be distinguished (using in explication a prohibitory predicate):[71] (*a*) *Must-rules*—the act is never to be done, could not conceivably be justified (cf. religious notion of utterly damning, moral notion of infinite obligation as contrasted with finite, ordinary notion of utterly unforgiveable act). (*b*) *Always-rules*—this act should never be done in any case that will actually occur; although conceivable conditions might justify it, they will not take place. (These are acts of finite but very high obligation.) (*c*) *Phase rules* (or *break-only-with-regret rules*)—the act is to be regarded as diminishing the result in any value-reckoning; it is a negative weight but may be outweighed; even when outweighed the drag of its weight remains.[72] (*d*) *For-the-most-part rules*—rough frequency with which the act turns out to be wrong. There are no doubt other types which a careful survey of different moralities would discover; e.g., Linton calls attention to "Do not do this, but if you do it, go about it in this fashion."[73]

The many notions fused in the moral law concept and the diversity of forms point up the scientific issues underlying the occurrence of given organization patterns. We can then investigate what degrees of determinateness in the field of moral phenomena are required to support what kind of rule-type and, in an evaluative enterprise such as working out standards for a given field, pose the question which should be strict standards, which flexible, which partial, and so on.

(iii) Since theories of virtue and vice are centrally anchored to the phenomena of character and character-formation, organizing schemes in this area reflect directly the underlying EP's. Plato interprets virtues as the qualities of different parts of the soul, fitting the whole for successful pursuit of life's quest. The medieval religious outlook sees them as qualities of spirit unified by acceptance or turning away from divine will. A Hobbesian materialism sees them as traits reasonably directed toward maintenance of peace and order. A Benthamite utilitarianism sees them as unified in a prudential pursuit of the general happiness. A Marxian materialism sees them as the character-types fashioned by the productive processes and relations of man's material life and selectively reinforced in the light of dominant class needs. A Nietzschean voluntarism sees them as the will asserting or thwarting itself in a heroic or Hebraic-Christian configuration, respectively. A Millian liberalism sees them as forms of reasonable individual initiative in the human pursuit of general happiness. While an objective idealism finds the unity of virtue in the historical growth of the moral ideal, a Deweyan emphasis on the role of ethical reflection in mediating change sees virtues as methodological qualities in the successful pursuit of human interests. In contemporary scientific studies of virtues the influence of underlying scientific theory on organizing ideas is explicit. Behaviorist trends see virtues as habits established in manifold areas, endowed with no central unity, bound to specific contexts of application and sanctions. Phenomenologically inclined psychologies look for phenomenal invariants, general essences, to differentiate, for example, sympathy from pity and mutual respect from mutual advantage. Psychoanalytical psychologies look for character-types reflecting the growth of personality and the typical deviations and distortions at different stages. Culturally oriented psychologies look for personality patterns (each with its specific virtue-set) corresponding to dominant culture patterns transmitted in socialization (educational) processes. In all these cases the same lesson is reached for the virtue field as for the previous ones: just as 'means,' 'ends,' 'moral law,' and the rest will be found to take their shape from the theories of human goal-seeking, appetition, regulative volition, and so forth, and the conditions of their exercise, so categories for understanding virtues express the answers given in the psychology of personality and its development.

(iv) With the development of twentieth-century linguistic-logical analysis, many of the formal framework questions have been brought into sharp focus. Two tendencies have been found in this analysis. One is to carry out as formal a logical reconstruction as possible: to formulate explicitly syntactic and semantic rules and, where possible, to provide a deductive system adequate to the field. The other analyzes actual linguistic usages, not to develop powerful logical instruments, but rather to render linguistic habits explicit and remove paradoxical cases.[74] In the carrying-out of both types of analysis, a point is reached where alternatives have to be faced. The formalist may develop different whole systems of deontic logic, and the ethical theorist will have to decide which to use or which is most adequate for ethical inquiries. Similarly, the informalist may be faced with conflicting patterns of linguistic usage and so have to decide which is correct or which is more serviceable in the light of specific aims.[75] At this point, questions of existing language-habits or conceptual patterns, questions of aims or purposes in the specific field, and factual or scientific questions about the materials in the field arise. How relevant is the third consideration for decision in formal framework questions? The brief illustrations that follow are from issues that have at various times agitated ethical theory.

a) One problem concerns the subject-type to be permitted in sentences with ethical predicates, e.g., of the form '*X* is right (wrong),' '*X* is good (bad).' If we line up some alternative answers we find:

Only a will-act or decision can be right or wrong (Kant, Schlick).

'Right' should be used only of behavioral acts, not act plus motive (W. D. Ross).

Moral judgments are not really passed on intentions but on persons (Westermarck).

Any kind of thing can be good.

Only a conscious experience can be good; everything else can be only a source or cause of good.[76]

How can we decide, whether in determining the syntax of ethical sentences or in determining correct use, which path to follow? The kinds of arguments used throw some light on the issues. Schlick thinks only decisions deserve the name of conduct because they are in fact the direction-molders of human life. Presumably, if this assumption turned out false—if decisions simply registered trends that had already stabilized themselves —this change in the psychological theory of the role of the will would remove the ground for this selection of subject-type. Westermarck is explicitly appealing to a psychology of the emotions in moral judgment. Those who insist on experiences as sole possible subject for 'good' may already have in mind a specific interpretation like 'pleasure' or else be operating in general with an EP in which value is 'subjective.' Ross offers his view as a proposal—that 'right' be confined to behavioral *acts* and that 'morally good' be used for *actions* (differentiated as an act done for a certain motive)[77]—for two chief reasons. First, good consequences and good motives do not always co-incide; presumably, the refinement would spare us the paradox of saying "He did a wrong act" when a man helped another from a selfish motive such as reward. (But this introduces the possibility of other paradoxical-sounding expressions, such as "It is morally good to do such-and-such a wrong act.") Second, Ross finds it difficult to allow that it is my duty to have a given motive or do an act from a given motive, since it cannot be my duty to do what is not in my power. Hence the psychological assumption that motives are not within our power is one of the grounds in his stipulation of subject-type. Without entering into the psychological question of such control, and the parallel problem whether men can be told to control their feelings— which carries us into the whole character of repression and its effects in human life—it is clear that the decision on appropriate subject-type here becomes to a great extent a function of the state of psychological knowledge and control.

b) Another type of problem concerns the mutual relations of ethical terms. Take the perennial issue of the relation of the right and the good. Here again, no matter how we approach it, we end up with alternative accounts to which we may affix labels:

The *goodist* structure: 'right' is to be defined as 'productive of the greatest good.'

The *rightist* structure: 'good' is to be defined in terms of 'right'; for example, to call anything good is to say that it is what a good man would choose, and a good man is one who does what is right.

The *co-ordinate* structure: 'right' and 'good' are separate and co-ordinate terms; propositions relating them are synthetic; any interpretations for them are distinct.[78]

The formulation "What is the relation of the right and the good" telescopes several different types of inquiry, which can be untangled by looking at the sort of considerations deemed relevant to decision. It cannot be a purely analytic problem if the three structures are all found in use, and if—as seems likely—each can be extended to cover the whole field of preanalytic phenomena.

Descriptive issues enter either directly or as dominating considerations in an evaluative proposal. For example, the great strength of the co-ordinate structure probably comes from men's recognition that this corresponds to the picture of their phenomenal field: they experience consciously a conflict of duty and interest rather than a conflict of two interests. Those who seek to relate the right and the good may question this phenomenological picture either by presenting a different one or by questioning its ultimacy in terms of an explanatory account. G. E. Moore would say that what he means by 'right' is what is productive of the greatest good; or the same goodist formulation is cast in a linguistic mold by Nowell-Smith when he takes 'good' to be expressive of a pro-attitude and argues that "pro-words are logically prior to deontological words" in the sense that "they form part of the contextual background in which alone deontological words can be understood, while the reverse is not the case."[79] If there really are two different phenomenological pictures, then the issue between them has to shift to an *explanatory* account. Each side will offer its own scientific (probably psychological) theory. For example, Westermarck argues that the concept of duty can never be reduced to that of goodness, because the former springs from the emotion of moral disapproval while the latter springs from that of moral approval, and—this is the crucial point—he finds these to be psychologically quite different in their sources.[80] On the other hand, Utilitarianism seems to have a complex logic supporting its goodist structure. The full theory of human psychology and social and historical development, it is believed, will show obligation experiences in all their detail to be a function of the pursuit of aims by men, under specific conditions of development, with more important paths of striving becoming associated with more demanding phenomenal qualities and more stringent feelings. Where the obligation experiences go off on their own in contrast to value experiences, it may be construed as a distortion, or psychological lag, or lack of clarity about aims, as well as (perhaps most importantly) the conflict of aims in different groups on the historical scene. Therefore, the goodist structure is taken to represent the most reasonable long-range interpretation of the dynamics of moral phenomena.

Clearly, these issues express the conflict of different EP's, and it is the answer to the scientific questions which is prerequisite to the theoretical decision. When the issue is eventually posed as an *evaluative* one, questions of the purposes that ethical theory and structures of this sort serve will also, of course, be relevant.

c) As a third illustration, take the question of the correct logical form for ethical sentences. In the last few decades, we find such views as:

"Killing is wrong" is really a way of saying "Thou shalt not kill."

"Stealing is wrong" is to be interpreted as "Stealing!!!" where the exclamation marks indicate an expression of horror.

"Friendship is good" is to be interpreted as "Would that everyone desired friendship."[81]

These were only a beginning; they were followed by the full force of emotive theory and a variety of increasingly sophisticated formulations.[82] In contemporary theory the range of these alternatives tends to be worked on in two ways. The formalists construct logical systems, with each of these notions providing the central undefined concept. Thus we can have a logic of imperatives, of optatives, of permissives, and so forth—not excluding a logic using the indicative form with such a concept as 'better' in a central position.[83] The other path is to map functional contexts corresponding to each particular alternative—indicatives instruct or inform, imperatives affect the will, optatives influence aspiration, and so on. Whether the work is done in the formalist or the informalist mode, the outcome is a set of alternatives, and the issue of decision once again has to be located as either analytic, descriptive, causal-explanatory, or evaluative.

Once again, the analytic approach discovers and works out alternatives but cannot decide among them without reference to some aims or some factual assumptions. To say that indicative formulations are really an unclear way of expressing commands is to claim that contexts of indicative ethical use will in fact be found to be command contexts. Or else the analysis may be making a methodological proposal that ethical utterances be construed as commands because this relieves us of having to regard ethical utterances as true or false. (Defenders of the ethical indicative would regard this as begging the question, and some logicians have even tried to construe imperatives so that truth values could be assigned to them.)[84]

Inquiry is compelled therefore to go into descriptive and explanatory questions, either to find existential interpretation for the central terms of formalist systems or to explore the human activities in the contexts selected as distinctive by the informalists. In either case we are carried into psychology—the nature of commanding and the authoritarian situation and the type of personality involved; the nature of wishes, idle and efficacious types; the place of intelligence and knowledge in motivation; the ways of influencing people and the respective force of subtle urging and providing insight; and so on. Similarly, we may be led into the social sciences—the actual moral codes in different peoples and cultures or subcultures and social strata; the extent to which they take the form of, so to speak, an indicative morality, an imperative morality, an optative morality, and so on; and the social, cultural, and historical conditions under which they develop.[85] Such scientific study would give us a wider basis of knowledge for assessing the different forms. We would know how far a given form of discourse—say, again, the imperative type—arises in or is sustained by authoritarian institutions and repressive psychological techniques, how far it is part of a special personality-structure, how far it successfully maintains that structure or touches off inner rebellion. We would also have educational generalizations about the comparative effects of commanding, or showing and predicting consequences, or cultivating responsibility through common modes of decision, or a host of comparable alternatives. As a consequence, our deferred methodological policy decision of the appropriate logical form for ethical sentences would represent the result of an evaluation in the light of assumed aims or purposes, on the basis of the results of scientific inquiry. Hypothetical alternatives may be envisaged. For example, if we found that always and everywhere morality embodied a dominant concern with repressing of otherwise irrepressible aggression, that the most successful technique of repression was peremptory will or command, that the imperative form of discourse touched off this technique—in short, if we took one of the traditional pessimistic views of human nature—then the imperative structure of ethical sentences would be both natural and represent sound methodological policy. If we took the more optimistic liberal assumptions about man, his deveopment and prospects, the imperativist form might very well represent an authoritarian vestige, and some aspirative form—optative or indicative, depending on the role we saw knowledge taking— a sounder policy. And so on. And decision among these views is ultimately a scientific question.

The three illustrations of formal framework questions and

their character, show how—though not in any one or uniform way—scientific considerations are relevant to decision no matter in what terms the decision problem is initially cast. The discussion of justification modes that follows continues this lesson in a kindred area.

20. Validation, Verification, Reasoning, Justification

These terms tend (with considerable shifting and uneasiness) to be used as follows: 'validation' in general for an exhibition of adequacy; 'verification' where the crucial appeal is to some form of experience; 'reasoning' for some legitimate form of inference; 'justification' in a blanket way for any acceptable direction of appeal under criticism. Let us use 'justification' as the generic term. The salient justification-candidates in ethics have been (i) intuition of a universal, (ii) perceptual disclosure of a particular, (iii) deductive derivation, (iv) inductive establishment, (v) normative (persuasive) success, and (vi) furnishing good-reasons.

(i) *Intuition of a universal.*—Intuition is generally mistrusted in contemporary philosophy as a mode of grasping truth. But intuition-claims have covered a variety of possibilities requiring separation. In ethics we find:

Stipulative definitions, such as "Courage is virtue with respect to fear and confidence." (If this is regarded as lexical definition, reporting usage, it is not, of course, intuitive.)

Abstract schema with blanks to be filled in. "Justice is giving every man his due," if not stipulative definition of 'justice,' is almost a sentential function, with 'due' a blank to be filled in by some apportionment-criterion.

Analytically true or false statements. "It is wrong to misuse social institutions," which Baier offers as part of absolute morality,[86] seems analytic, since the idea of 'misuse' already includes that of using wrongly or badly. (It may be an empirical invariant, if what it means is that every actual morality will be found to contain such a conception, to which it will invariantly assign a negative value.)

Theorems in insufficiently explicit systems. "To prefer a lesser good for one's self rather than a greater good for another is wrong," often urged as axiomatic, is usually derivative from certain definitions of 'good' and 'wrong' whose effect is to rule out the expression 'my good' and make all size estimates of good impersonal.[87]

Basic or abstract value-affirmations. Even so apparently methodological a proposed intuition as "Hereafter *as such* is to be regarded neither less nor more than Now"[88] in effect is recommending a rational attitude oriented to the totality of life, for which there are conceivable value alternatives. More specific basic intuitions turn out to be vehicles for special forms of life or social organization: e.g., a man is entitled to the full produce of his labor, or a man in distress ought to be helped. In the long run, basic value-axioms are assessed not by one's intuitive response but by the system of life they organize, which is a more complicated evaluative process.

Phenomenological reports. Some contemporary claims for intuitive moral universals of a lower order—e.g., "Cruelty is wrong"—insist that self-evidence characterizes the object of beholding; there is no appeal to a mode of intuiting as a process that of itself guarantees reliability.[89] In that case, self-evidence is best interpreted as a phenomenological property of the object in the field, like a man having an honest face or a sinister look. In this sense, 'intuitive' does not carry the connotation of true or self-justifying. Thus "Cruelty is wrong" as self-evident either would be a single-person datum concerning his phenomenological field or would involve the further claim that all men would give the same report for a similarly structured field. (Compare Karl Duncker's argument that the same valuation will be found for all phenomenal fields in which the meaning grasped is the same.)[90] This is no longer a simple intuitive inquiry but a transcultural one in which phenomenological reports play only the role of verifying observations—in short, a scientific inquiry.

Whatever interpretation is taken of a given intuition-claim, it therefore requires further justification along one or another of the lines indicated.

(ii) *Perceptual disclosure of a particular.*—A long tradition from Aristotle's day takes a singular moral judgment to be certified in a kind of perceptual act, enabling it thereafter to serve as a hard datum in verifying general propositions. Here again we find several types:

Immediate affective *acts* or *states* as moral phenomena: for example, being pleased or feeling obliged. As acts or states, they require conceptualization and interpretation. Hence they are not hard data but can in practice be refined into serviceable pointer-readings.

Immediate *reports* about the occurrence of affective acts or states. These present conceptualization and so are in principle corrigible. For example, a report that a person is pleased opens the way to considering whether it is pure pleasure or relief from anxiety or any of a variety of qualities that a developed psychology might find. Hence whether it is a simple pleasure-report or a more complex obligation report or commitment report or any other phenomenological report, it becomes a reliable cognitive disclosure only to the extent that its significance is determinable by reference to scientific theory.

A synthesis in the particular situation of complex factors apparently not capable of prior analytic separation and compounding: a man grasps directly in a complex situation where his duty lies. This is perhaps the most typical and also most useful of the claims for intuitive particular disclosure. Here the procedure is less like that of scientific verification by hard data than that of application involving selective sense of relevance and synthesis of claims arising from different features, together with an ability not to overlook what may be pertinent. This operation, like that of the engineer or the judge, is necessary; but justification is another matter, lying in unraveling the strands and assessing their weight.

(iii) *Deductive derivation.*—How far ethical statements are justified by reasoning from an established theoretical system or some fragment thereof depends in the long run on how far such systems come to be established. Some deduction there will always be; on the other hand, the rationalist's dream of a complete code deductively formulated, is no doubt just a dream. How far ethics can move in such a direction depends on the objective character of the human field and the order it actually proves to have. If it supports generalizations, and stable definitions with well-demarcated modes of application, if the indeterminacy points in choosing among rules, in interpreting them, are not too great, then some measure of useful deductive form will be achievable.

(iv) *Inductive establishment.*—This again depends on a certain degree of stability and determinacy in the field. That there be existential links for ethical terms is a necessary condition; but that the material to which the link goes have some order and that it be discoverable are further conditions. Thus the use of induction in a utilitarian morality depends on there being reliable regularities in what brings people pleasure. Similarly, a morality that speaks of resolving problems or satisfying needs can be inductive only to the extent to which there is dependable achievable knowledge of needs and problems and modes of satisfaction.

(v) *Normative success.*—At a certain point, one segment of contemporary ethical theory lost faith in the possibilities of any of the above methods of ethical justification. The emotive theory embodied this shift. It gave ethical terms the specific existential linkage of reference to a context of mutual persuasive effort in cases of disagreement. Ethical reasoning was construed not as a logical (deductive or inductive) relation between statements (premises and conclusion) but as a causal relation between beliefs and attitudes as events. Inductive evidence could be used to establish beliefs, but the relation between the belief and the attitude as causal was subject to individual differences. Therefore the concept of validity in ethics is in fact abandoned.[91] The realities of the situation are pressure, influence, persuasion. A method of justification is itself an object of advocacy, specifying a pattern of transition from factual premises to ethical conclusions.

In assuming no dependable regularity in cognitive-emotional or cognitive-volitional relations, this view anticipates a particular outcome of the study of the human field.

(vi) *Furnishing good-reasons.*—A widespread current approach suggests that even in the absence of fixed *general* criteria for going from factual premises to ethical conclusions, ordinary language supplies a validation concept for ethics. There are in different contexts acknowledged patterns of good reasons for deciding or acting in given ways. There are informal or "unscheduled" logics or implicit rules of what is relevant and what is not.[92] Ethical terms are interpreted practically by reference to the contexts of choosing, deciding, advising, etc. Validity consists in conforming to the contextual rules. A good reason for giving someone a book is that you borrowed it or promised to return it. It is a good reason for obeying a moral rule that it is part of the code, for changing an item in the code that it will reduce human suffering, and so on.[93]

In this mode of analysis some regard reasons as reasons for doing or deciding or being under obligation rather than as reasons for believing or statements logically supporting (deductively or inductively) other (ethical) statements. Even in analyzing statements, there is sometimes a considerable extension of the concept of reasons, so that assertions relating properties are translated so as to specify reasons. Instead of saying "Promises ought to be kept," it will be said: "The fact that I promised is a good reason for my doing the act (or for my being under obligation to do it)." For example, Baier translates "What shall I do?" and "What ought I to do?" into "What is the best thing to do?" and this in turn means 'the course supported by the best reasons.'[94] This would be helpful if the analysis of 'the best reasons' constituted an improvement on that of 'good' or 'ought.' But we are told that we act "in accordance with what we take to be the best reasons . . . because we *want* to follow the best reasons";[95] "We mean by the word 'reason' something that can make us do things."[96] But in the end "The criteria of 'best course of action' are linked with what we mean by 'the good life.' " And, "Our very purpose in 'playing the reasoning game' is to maximize satisfactions and minimize frustrations."[97]

There is a kind of arbitrariness in translating everything into the language of good reasons, because no new mode of justification is introduced thereby. The mode of justification is either that the language habits have that pattern (or contextual logic) or else simply that it is intuitively clear that making a promise is a good reason for keeping it. In some respects the situation is parallel to translating a material general statement in science into a material principle of inference. And perhaps the reason is the same—it stresses the conventional role rather than the empirical character. But it must not be forgotten that this should be done only to a well-established statement.

If, however, the immediacy or intuitive character of the apprehension of good reasons is stressed, then the whole approach can be most clearly seen as a variant of a phenomenological analysis. The good-reasons relation between a fact and an action is a kind of vector in the field when both are contemplated or "beheld" side by side. But if so, then we have argued above, this is not a mode of justification but a datum or an observation-statement which may serve to confirm or establish some statement or theory relating phenomenological to non-phenomenological data.[98]

Such an interpretation of the good-reasons relation, especially as applied to the relation of beliefs and acts, rather than statements, helps bring some order into the uses of 'reasons.' We can then see that the concept of reasons varies with the method of inquiry. If it is analytic method, the relations are logical, and reasons are premises. If it is descriptive, the relations between reasons and what they are reasons for is phenomenological. If it is causal, then reasons are provided when we have given a causal explanation. If it is evaluative method, reasons are furnished for given acts when we have indicated the ends or purposes that the acts serve. In short, the very concept of reasons has a contextual variability depending on the method of inquiry.

In general, the different processes considered in this section rest more profoundly than is commonly realized on the extent of the determinacy of the field. Intuition of a universal is practically applicable where there is considerable stability in phenomenological field characteristics. Perceptual disclosure of particulars will be appealed to where complexity is great and the disorder high, so that judgment seems practical art rather than calculated conclusion. Deductive derivation, as we saw, has extended applicability only in the conditions of greatest stability and relative steadiness of classification. Inductive establishment presupposes a moderate degree of determinacy sufficient to support some verification. Normative success as typical analysis of justification betokens that field condition in which the context of ethical discourse is clear but few dependable features characterize the relations in the field. The good-reasons approach sees fragmentary contexts of relative stability. It is not our purpose to decide here what is the state of the field but to note that *the various justification processes and theories about them reflect such estimates*. This is the sense in which we suggested at the outset that methodological policy decisions embodied some anticipation of scientific results.

21. Application and Evaluative Processes

As we move into the theory of application and evaluative processes, the specificity of situations becomes greater, conditions of special fields are added to the general contours of the human field, decision is directed from generalized principles to particular concrete situations culminating with the here-and-now. Here is where different bodies of factual knowledge converge, skill and insight combine with "know-how," informal procedural principles of relevance and selection begin to loom large. Here the significance of factual materials is unavoidable. But here, too, questions of complexity and subtle differences in evaluation have greater force. Weights and measures so often fail to be at hand that scientific aid seems out of sight. Although we cannot enter into a systematic examination of application theory, a few extended suggestions may be offered by way of comment.

(i) There is a pervasive evaluative process going on at the grass roots in human life, constantly raising reference-points, building them into evaluative criteria, fashioning these into partial standards, and greeting a more systematized morality as a guide in an ongoing process. Certainly, there is no dearth of reference-points, and we often become conscious of them when they are already functioning as criteria. You stretch your hand out for an apple and are already looking for one that is firm and red-cheeked; the apple industry has to go further and articulate a whole theory of grading.[99] You want a drink of water; do you want it enough to interrupt what you are doing? Even in so simple a desire you are faced with the standpoint of its strength and the collateral effects of satisfying it; a social enterprise like education has to work out careful standards of discipline and permissiveness for all stages in the growth of the young. Large-scale social processes—sorting and quality measurement in industry, legal regulation in realms as diverse as health conditions in ocupations, or the marks of substandard housing, or the fair value of a public utility as a basis for rate adjustment, or the meaning of reasonable caution or negligence in automobile driving, and so on—show clearly the interplay of empirical, technical, theoretical, and ideal or purposive elements that enter into standard-formation.

Ethical theory (with the exception of general value theory)[100] has not sufficiently explored the richness and variety of possible criteria in concrete evaluative processes within the moral domain. Consider, for example, criteria that arise when a man or a group not merely contemplates a proposed means or end but deliberates whether *to adopt* it. For means, one might suggest: Will the means if acted on produce the end (its *effectuality*)? What is the quality of its performance (*efficiency*)? How does the use of the means affect collateral ends (*constructiveness* or *destructiveness*)? If there is a considerable investment of energy and resources in providing the means, for what other ends can they be used (their *multivalence*)? What is the *liberating-power* of alternative means to the same end? To what human needs or drives may the means-activity itself give expression (its *expressiveness*)? How satisfying or enjoyable is the means-activity itself (its *luster*)? How far will a particular means if utilized tend to become an end in its own right (its *"telicity"*)? What is the resultant *cost* of employing a particular means, in terms of the disvalue of the means-activity, the end-concomitants, and the consequences? For ends: How *attractive* is the end envisaged by itself? How far capable of occurring without means-components (its *purity*)? How long-lasting (its *permanence*)? In relation to other goals, what support does the occurrence of one end give to others (*constructiveness* or *destructiveness*)? What is its *area* in the field of endeavor of the given person or group? In relation to means, what is its *attainability* and its *cost*? In relation to the personal and social economy, what is the strength of the underlying drives and problems to which it is addressed (its *depth*)? With respect to these, what is its *role*? In the light of its role, does it prove to be spurious (e.g., rationalization) or authentic (insightful and realistic in grounding), that is, how *genuine* is it? What is the degree of *satisfaction* that it brings?

Criteria for evaluating ideals will overlap with those for ends but may have novel elements corresponding to the nature and role of the ideal as a human phenomenon.[101] In the obligation-family a fresh set of reference-points may prove relevant; for example, *stringency* plays the part here that attractiveness plays in ends. Similarly, virtues can be estimated from the point of view of *utility* as well as *attractiveness*, and, once they are set in a psychological and social context, there will be criteria of psychological *maturity* of a given virtue, standpoints provided by the social forms for which the virtues act as *girders*. And so on. It is also possible that reference-points may rise initially from scientific exploration. For example, Erikson, with an eye on the development from babyhood to maturity, suggests that criteria may be charted for decisive ego-victories at given points.[102] This suggests standpoints reflecting an acceptable theory of the development of the self. These might in turn be reflected in a revision of criteria of obligation (for example, a clearer picture of sincerity as against rigidity in the sense of duty) or of virtue (for example, in giving body to the idea of mature virtues as against reaction-formation types). And the development of a reliable social science has comparable effects. The route of such standard-formation is from the science through the EP to the development of criteria.

(ii) At the same time that there is this creative process of building up evaluative criteria from below, and consolidating them into standards, there is also a parallel process going on higher up. This consists in gearing the morality as a more or less stabilized system, to the tasks of application in a determinate setting. Elements of the morality—whether significant content, methods or procedures, special features manifested in special persons, and so on—move into a guiding role for evaluative decision because they are able to exercise the office of standards in the existing situation. They are selected for this in a way not unlike that in which natural entities in virtue of their regularity of process come to function as clocks, or special commodities like grain or gold come to function as media of exchange and modes of reckoning value. A comparative examination of the kinds of standards that emerge as a morality prepares for action adds to our understanding of evaluative processes.

Most prominent perhaps is the *paramount ideal* type. Ends of generality and abstractness that organize a great area of effort are prone to grow into ideals and serve as standards in appraisal. Important feelings, fundamental drives, and obligation-relations of broad scope may also function in this way. Happiness, harmony, justice, and the satisfaction of one's whole nature has each at times stood out as a key standard. Virtues too have sometimes grown into the role of central ideal, so that the very notion of "the moral ideal" came to mean a standard of character-development. These are being considered as standards rather

74

than as ends or as the content of "the good life" when they are elevated to a supreme position for guiding conduct. In some cases, even a fairly concrete end may gain in scope and influence and come to furnish a standard of this sort for particular lives or particular groups.

Method standards also occur, reflecting the methodological properties of different EP's. Bentham will want to know if a decision has carried out a rational calculation; Kant will look to its conscientiousness; Dewey will look to its reflective character; and so on.

Individual models furnish functioning standards more widely than we are likely to think. The man of practical wisdom, the exalted leader, the impartial spectator, the saint, are familiar examples. To sense the spirit of his action, although not able to describe it adequately or reduce it to rule, provides a kind of dynamic sensitivity. It is not only persons who constitute models. As Parker points out, every striking event—a deep satisfaction or great deed, or even a first experience—may rise in memory to provide a touchstone for anything resembling it.[103] It resembles the clear case in definitional theory where a formal definition has not been achieved.

Jurisdictional standards tend to be overlooked in spite of their great importance. One issue is declared a matter of duty; another, a matter of taste. This is an academic matter; that one calls for action. Here we have a compromise situation; there, only a question of expediency; there, strictly a matter of principle. It is not always easy to see what criteria are being employed. Sometimes they are rough judgments of urgency, of importance, of complexity calling for deliberation, and so on. And there are hosts of relevance criteria in different areas of content that may predetermine the kind of result appraisal will bring. In some critical cases the pivotal point may lie in calling these relevance criteria into account for their own appraisal on some other standard.

One of the least recognized standard-types, although in practice it may be one of the most important, is the *indispensable-means* type. A means of wide scope and critical importance because many ends require it comes into central attention and functions as a widespread standard. The need for peace or the need for industrialization in a great part of the world becomes such a standard. In the extreme case its logical form may be simply as follows: Let every party concerned specify his own ends without even revealing them. Then the general proposition is asserted that, for all or almost all such ends actually held, it will be found that they cannot in fact be achieved without the common means. Hence the means becomes the immediately applicable standard. (Hobbes on peace and security is the obvious historical instance.) Braithwaite makes a similar point in another context: "In this Kingdom are many mansions. It is more reasonable to seek to enter this Kingdom by the only known modes of entry than to postpone the attempt until assured as to which, if any, of the mansions is the ultimate end of the quest."[104]

Finally, we refer to a whole host of *totality* standards. These are characterized by reference to a whole which is to provide the basic standard. It may be a whole life of the individual[105] or the idea of the good of the whole community, as in many ethical theories, both individualistic and organic. Or it may be a whole-history ideal, such as the growth of freedom of mankind collectively.

(iii) The outcome of criteria development from below and standard-formation from above may be a greater stability in moral judgment than the complexity of the processes might suggest. How stable it is depends—as in the theoretical problems of morality discussed earlier—on the determinateness of the field. And this, which also is basic to the policy decision how far to employ scientific method in application problems, cannot be settled a priori. Here, again, the results of a scientific examination of the field are prior to the policy decision about method. How necessary this independent examination is can be suggested by showing—paradoxical as it may sound—a few types of cases in which the standards in application may prove *more dependable* than the morality that is being applied.

The clearest illustration is the indispensable-means standard just considered among the inventory of types. If its achievement is a prolonged affair and affects a great part of life in a given epoch, then it can operate as a dependable standard in spite of the disagreement about ends.

Sometimes a moral standard may be stronger than the general moral principle on which it rests. Take, for example, the standard of honesty in the scientific profession—that is, of a scientist in his scientific work. The general moral principle of honesty as a virtue has a long history. It merges sometimes with wider virtues of integrity in a human being and authenticity in interpersonal relations. It has hard going in some institutional fields—for example, in the business world, where "Caveat emptor" could long reign supreme, or in political fields, where international morality countenances espionage and counterespionage. There has been some progress in the justification of honesty as a basic moral virtue, especially in the light of recent psychological advances in the understanding of the scope of self-deceit and the importance of insight in human well-being; and it is likely that major agreement could be achieved on its acceptance as a phase-rule. There is still, however, a large gap between honesty as a moral virtue and honesty as part of the functioning standard in many areas of application. But, as a standard in scientific work in the modern world, little discussion is needed, in the light of the role, goals, and co-operative procedure of inquiry in the field. The standard does not answer whether men should become scientists, but it is scarcely shakable that if you become one you are obligated to follow the standard of honesty in your work.

Another context is the convergence of different underlying moralities on a given conclusion within a limited range: Christian, Utilitarian, Marxian moralities all include the idea that racial discrimination is wrong. Of course, the moralities converging here might yield varying conclusions beyond the ordinary range, just as two mathematical formulas may yield the same finite set over a given range in the series and diverge beyond, or a physical law maintain a given form only within certain limiting conditions. But if the standard is set for the given range, and it is probable that the problems to be faced will arise within that range, then the standard may turn out to be supported by all conflicting moralities. It may therefore be more dependable in its domain than any one morality in general.

Another possibility is that a standard may rest on the applications of those parts of moralities that are invariant—commonly recognized goods, or isolated obligations—or it may rest in some cases on compromise portions of a morality, as where the standard calls for the act while morality argues about the motive, for example, in what often appears as standards of "common decency" in neighborly relations.

Fuller logical exploration is needed of the types of specific relations between elements of the morality and elements of the standards that apply it. But it may be possible to construct a consolidated standard for a given age cutting across major cultural and ethical theory differences. Such an integrated standard would have a role in evaluative processes similar to that played in factual determinations by the body of existent scientific knowledge of a given period. This attempt has been outlined elsewhere by the present writer in a concept of the *valuational base,* as an interlocking structure of human knowledge and human striving which embodies the best available knowledge of man's aspirations and conditions.[106] Its constituents would be fundamental human needs, perennial aspirations and major goals in their specific sociohistorical forms, central necessary conditions, and critical contingent factors. And a study of the way in which items would be included or excluded from this base showed clearly the penetrating role that findings of the psychological and social sciences might play in formulating the standard.

IV. Decision, Freedom, and Responsibility

22. Evaluative Processes in Unstructured Situations

Underlying the treatment of the place of science in ethics has been the hypothesis that the human field has a greater degree of *determinacy* than we are prone to assign to it and that problems in human life which generate and support standards and in which standards are utilized present themselves as more or less *structured* because the field has sufficient order in terms of continuing conditions, continuing aims, and modes of interaction. Let us differentiate the terms, thinking of the field as the scientist studies it as *determinate,* and the field as it appears phenomenologically in the view of the person making decisions as *structured.* Now if the determinacy of a field either because of too great a complexity or too great an instability falls below a certain point, the very problems seem to lose their structure. Established standards no longer seem to apply, and new standards seem incapable of establishment. There is a problem, but no definitive analysis even of what it is. What is the character of evaluative processes at such points?

There are probably many different types of ill-structured fields, and evaluative processes will take different shape depending on the factors that give rise to complexity and instability. Let us briefly survey sample types along a range from the highly structured to the practically unstructured.

In the "normal" highly structured types there are well-established standards and clear criteria of jurisdiction. Problems are treated by bringing to bear existent aims or existent knowledge to broaden or sharpen the standard, or by otherwise "harnessing" the indeterminacy.

A complex area with considerable variability may be given the jurisdictional mark of "to be settled according to taste." This is often misunderstood as the absence of standards, whereas it may be an extension of a libertarian ideal to broaden the area of individual choice; this is obvious in the area of food preferences, less obvious in that of reading preferences, and becomes a controversial issue in the modern world in the advocacy, for example, of planned parenthood as expressive of parental choice. Again, in dealing with fringe-areas, the problem is felt less as one of lack of standards than as a specific question of determining the jurisdiction of existent standards. For example, there may be doubt whether an issue is a professional matter and so governed by the stricter rules of that profession, or a personal matter and so more open to subjective considerations provided they are "sincere," or a legal matter subject to established law. In general, however complex the task, the evaluative process is felt as one of *applying* existent standards; although recognizing the role of interstitial "legislation," such a view never feels it as wholly unbound or arbitrary.

Where the indeterminacy arises from complexity of conditions, or from changing conditions, but not from changing aims, the field of decision becomes less specifically structured, but by no means wholly unstructured. The shift may be from content-standards to model-standards or method-standards. These may vary from assignment of fixed or authoritative locus of decision —papal infallibility, the decision of a controlling party, or the procedurally determined expression of popular will—to a general philosophy of method (e.g., Dewey's reorientation of ethics).[107] Up to a point such a substitution of decision-method for content may hold. But if the indeterminacy involves large components of conflicting or changing aims, the very methods as standards may become a battleground.

Where the indeterminacy is clearly grounded in conflict or change of aims, the decision situation becomes structured in the broadest value terms. Virtue-standards may come to the fore, and, as in late Roman Stoicism with its stress on the precariousness of life and the folly of committing one's self to specific aims and hopes, the fixed point of decision becomes the maintenance of personal integrity. Or else most generalized totality-standards are invoked; one is not told what to decide but simply to remember that one is deciding for one's whole life or for all mankind.

The end of the range is found in the extreme situation where the very aims of men are regarded as indeterminate and the conditions of life as utterly precarious. Here the situation of decision looms as central and almost solitary. Because of the great influence of this structure in contemporary ethics, it is important to understand its EP. When it speaks of decision, it is thinking not of the slow change in standards—as the American people changed their standard of responsibility for unemployment over the period of the Great Depression—or of cumulative changes in an individual's aims over decades, but of the lone individual in the here-and-now; it extends the category of *decision* to cover practically every turn that a man can make. (And especially since responsibility is tied up with decision, it has momentous ethical consequences.) But in the very extension, any support for decision is itself removed. If you follow established standards, it is an implicit decision to follow them; if you apply a principle, it is a fresh decision—a resubscription or a readoption of the principle in the very act. Hence decision itself, devouring any possible basis on which it might rest, becomes seen as pure act. It becomes the absolutely cut-off (free) evaluative process, which we may call *the Sartrean leap*.[108] In more analytic fashion the same point is embodied in current prescriptivist emphases in ethics that insist on reformulating the meaning of ethical terms by reference to the first-person or agent's present decisional situation.

Putting aside metaphysical or analytic dogmas, we can extract from such a perspective a number of insights about evaluative processes in unstructured situations.

(i) There is indeed a decisional element implicit in all deliberative use of standards, even in what seems to be automatic subsumption of a situation under a rule, for there is a jurisdictional assumption in turning it over to this rule. Probing for this decisional element has been everywhere useful in laying bare value assumptions—particularly in the philosophy of law, where the deductive model of interpreting and applying pre-existent law often had marked ideological uses. The complete mapping of the components of standard-application must always include the value assumptions, even though in the areas under investigation they may be practically negligible *because they are constant and deeply rooted.*

(ii) It is, however, by no means clear that such an unearthing of value assumptions means that they have to be reviewed. We have to distinguish the cases where bringing to consciousness means *bringing into question* from those in which it means *bringing into recognition.* The thoroughly legitimate inference that the possible scope of evaluative processes is endless and that nothing is exempt should not be confused with the assumption that consciousness always has a subverting role and that where it does not it is because there has been some fresh perpetuating volition. This is the issue we may epitomize as whether the Socratic quest for knowing oneself is really *finding* or *creating*. The present point is that there is no *general* answer. Sometimes it is recognizing what we are like and what commitments we have. Sometimes it is altering in the act of discovering. Exploration of these processes is best done by using 'finding' and 'creating' as empirical categories for psychological research rather than giving a valuational cast to the phenomena under investigation by some blanket appelation. This means probing for further knowledge about the relation of consciousness and volition, the extent of permanent unavoidable needs that play a constitutive role in the self, the influence of different conditions in motivational processes, perennial sources of internal conflict, and so on.

(iii) In the extreme case there might seem to remain at least one structured feature in any decision situation—the initiating question reduced simply to "What shall I do?" still means that *something is to be done*. If there is no remaining sense of aims, conditions, determinate dangers, definite frustrations, then its structural residue is at least a call for help. Below such minima, its phenomenological structure as a decision situation vanishes but its status as a phenomenon-to-be-understood still remains, with all its qualities of suffering and anguish. It is philosophical —especially existentialist—attempts to interpret such phenomena as in some sense furnishing a profound truth about the human situation and the nature of ethics that makes it important to show clearly the opposing avenues of theoretical solution. And these, it becomes very evident, bring us back to issues of underlying EP's.

The essence of the scientific analysis here is to insist that phenomenological lack of structure need not mean actual indeterminacy in the situation. The field conditions for the lack may be quite determinate. Psychological accounts of frustration in unstructured situations show that there are definite aims that are being blocked. Sociological accounts of anomie and its consequences do not deny that there are definite social and individual needs that have become entangled and channeled in such a way as to produce confusion in the consciousness of standards. In revolutionary situations in history, new social patterns have already formed even while consciousness is cast in terms of the older breakdown. So too in such personal experiences as dramatic conversion or pervasive object-less anxiety or dread. There is no need to take the one as a leap by a bare self behind phenomena or the other as a special metaphysical indicator. There are scientific problems enough in their study, but they are specific problems of discerning accumulating forces, the conditions of qualitative change, and so forth.

The essence of the opposite view is that the phenomenological feature of lack of structure in decision situations is central because it provides points at which we can break through the local blinders of ordinary situations. Wisdom in ethics comes from concentrating on the marginal indeterminacy precisely because it reveals to us the human predicament. Anguish and dread therefore have ontological significance and are not to be taken as generalized anxiety detached from its object or in some sense accumulated in human processes. In actual philosophical contexts, of course, such theses are tied up with specific EP conceptions, whether of ultimate subjectivity or the central reliability of phenomenological method, and so forth; but basic focus is on the phenomena and the issue of their interpretation.

Perhaps the scope of these issues and the place of science in their resolution can best be suggested by a central illustration—attitudes to death. At one extreme we find the Spinozistic maxim that the free man is not concerned with death; at the other we find the view that to see a situation in the light of death is the only way to be sure of authenticity in its understanding. Such opposing views might charge each other, respectively, with morbidity and perversity, or with escapism, bland optimism, and naïveté. The battleground extends over many a field. The existence or non-existence of basic anxiety, whether if it exists it represents the "rock bottom" of man or a frustration reaction, the nature and role of insight, the genuineness of man's desire for immortality or whether it is a reflex of an unsatisfied life, the possibilities of a happy life without sacrifice of depth, the fear of death and whether it is genuine or a realistic feeling of "lack of time" or a cover for unconscious ideas—all these and many more issues are involved in resolving the major theses. Obviously, in spite of the ideological use of every item in this inventory, the basic problem is part of the growing area of the psychological and cultural and historical study of human reactions. It may yet be possible to distinguish whose optimistic attitudes, having what qualities and arising under what conditions, are escape attitudes, and whose despair attitudes or "outsider" attitudes, with what qualities under what conditions, represent basically clinical symptoms, and to see how far propositions about death and non-being are really propositions about ways of facing life, or how far the reverse is basically true. If the broad outlines of a scientific philosophy today incline against the general thesis of the primacy of death, it is not an unphilosophical optimism but a refusal to accept a limited phenomenological picture of a given moment in the historical development of mankind under specific conditions as in itself authentic, without relating it to the broader accumulation of knowledge—biological, psychological, historical—that man has acquired of himself.

Our survey of the range of situations from the highly structured to the barely structured suggests that scientific results have a place throughout—that the controversies are analyzable into questions that are essentially scientific in type. Scientific method would seem to be applicable in actual evaluative processes according to the extent to which the situations are structured and in different degrees and with different types of standards according to the types of structuring. But beyond the limits of its applicability at any time, there arises the factual question whether the growth of scientific knowledge of man will increase the range of applicability of the method and also the methodological policy problem whether the conceptual apparatus of ethical theory should be increasingly geared toward such an extension. On the analysis carried out, such decision involves a reckoning of the dependable projection of the curve of increasing knowledge. It also raises questions, however, of limiting conditions, and in the tradition of ethical theory this focuses on the problem of freedom of the will.

23. Toward a Scientific Study of Freedom and Responsibility

In popular consciousness the problem of free will stands as a perennial road-block to the relation of science and ethics.[109] Science is tied to determinism, ethics demands responsibility, responsibility requires free will, free will is antithetical to determinism. How can we project a scientific study of this domain?

(i) *Location of phenomena and descriptive concepts.*—Two bodies of conceptualized phenomena can be found. One is the *voluntaristic group:* the sense of freedom, various shades of willingness from jumping at the opportunity, acting eagerly, to intending, trying; again, the sense of acting unwillingly or being coerced, intimidated, provoked, compelled, of slipping into an act, of acting against one's better judgment, or half-heartedly, of succumbing to temptation. The second is the *responsibility group:* praising, blaming, finding fault, feeling at fault; offering excuses, holding liable; holding responsible, feeling responsible, finding meritorious.

These groups of phenomena and concepts extend far beyond these samples. They need study in many terms: descriptive and analytic; first-personal contexts, third-personal, interpersonal; border-lines with other moral phenomena (e.g., feeling responsible and feeling guilty about, or holding liable and making claims on). The search for phenomena can fruitfully go into experiential and conceptual elements in religion and law as well as morality: expiation, grace, forgiveness, retribution, vengeance, punishment, crime, liability of animals and things, obsession, addiction, scapegoating. Linguistic analysis has to distinguish different uses so as to untangle the phenomena: e.g., 'fault' in the legal maxim "No liability without fault" refers to some criterion of voluntary behavior, whereas in 'finding at fault' it is often equivalent to 'holding responsible.'

The central aim is to identify clearly the specific phenomena involved and provide descriptive concepts, thus separating the phenomena from interpretations.

(ii) *Causal-explanatory study.*—Such inquiry finds many aids in the psychological and sociocultural sciences. Depth psychology furnishes explanatory concepts concerning origins and functioning of guilt-feeling, mechanisms of identification and internalization of standards, functioning of the superego and emergence of the ego-ideal, phenomena of self-accusation and mechanisms of lifting the load of guilt, and so on.[110] Social and historical explanations are directed to functions and changes in responsibility patterns and in voluntaristic patterns.

For example, what explains the abandonment of holding animals responsible, or perhaps why it lasted so long in some quarters? Under what conditions do societies establish distinctions between voluntary and involuntary commission of an offense and narrow down punishment to the former? Under what conditions do patterns of kin responsibility yield to individual responsibility? Why does Oedipus feel so profoundly guilty while Adam Smith can go so far as to speak of the "fallacious sense of guilt" of Oedipus and Jocasta?[111] Can we explain the growth of a concept of "liability without fault" in modern times in terms of the rise of large-scale industrial enterprises with statistically regular mishaps, so that bearing an economic burden becomes distinct from accepting guilt? And so on.

(iii) *Inadequacy of present theoretical categories.*—Even partial undertaking of such studies shows the difficulty of conceptualizing so vast a field under the two general concepts of freedom and responsibility. What credence then is to be given to general formulas relating the two groups—whether it be a philosophical slogan such as "Responsibility is meaningless without freedom" or a legal maxim such as "No liability without fault" (in the voluntaristic sense of 'fault')?

If the generalizations are synthetic statements, counter-instances can readily be found. In law fault becomes stretched from intentional act to negligent act to implicit undertaking to obligation implicit in the situation whether undertaken or not and even when unaware.[112] Similarly, the philosophical slogan runs counter to: religious views of predestination combined with responsibility, legal concepts of vicarious liability (e.g., of master for servant), the sense of unavoidable guilt in some ethical theories,[113] of original sin in Western religions, and so on. Now it may be possible to discount such an array of counter-instances by subtle analytic differentiation of some, rejection of others as incorrect, and so on. But this will involve working out a theory of the two groups of phenomena on the basis of careful study. Any attempt to short-cut this study by relying on an analytically or intuitively necessary connection runs the risk of exalting a possibly provincial consciousness into a universal law.

There are also evaluative questions in the relation of phenomena in the two groups. For example: If a man is provoked, should his punishment be less? If he enters a conspiracy half-heartedly, is his responsibility greater or less? How far should a man be held responsible for acts of others, and under what conditions? Is coercion always bad, and forgiveness always good (at least as phase rules)? It is therefore possible that the general rule that responsibility is meaningless without freedom, as well as its legal counterpart, may be operating evaluatively—for example, as the expression of the growth of modern individualism in the past few centuries, demanding that the individual be freed from obligations to which he has not given explicit assent.[114]

In fact, a critical inquiry should go further and seek the very basis of unity embodied in the theoretical categories rather than taking them intuitively. The general concept of responsibility has seemed to have a firmness of outline which the concept of free will has almost wholly lacked. It is oversimple to say that this is because the former is a practical concept and the latter theoretical. Both have practical and theoretical components. Perhaps the apparent clarity of the responsibility concept comes from the fact that it is anchored to a standardized family of rewards and punishments; the concept of voluntary action lacks this visible reminder. But we have no really adequate theory of why we punish or of how effectively punishment deters and blame hinders or what blaming does to the spirit of a human being (e.g., whether it arouses a constructive sense of shame or subtly threatens with a loss of love).[115] And while it is clear that voluntaristic phenomena depend for their understanding on a theory of the will, we do not yet have a full-fledged psychology which can give us an account of the role of thinking and feeling in the phenomena of will, and all these in relation to a conception of the self and its development and its ways of dealing with internal conflict. In spite of rapid contemporary psychological progress, in this theoretical area there is still more program than accomplishment.

Nor is it useful to go to the metaphysical tradition in the free-will problem. Historical study shows that the very meaning of the problem undergoes change with the underlying EP in which it is cast. Given supernaturalistic and dualistic stage-settings, free will is assigned as a central property to soul or mind, as against the causality or determinism operative in and among bodies. Given a deterministic materialism or naturalism, the very locus of the problem is found at other points; for example, the achievement under favorable causal conditions of a certain type of character, or the development of a certain type of consciousness. There is continuity in the sense that all are concerned with furnishing some theory of voluntaristic phenomena and relating them to responsibility phenomena. But the free-will problem is seen to be a derivative problem, in the sense of requiring for its understanding and formulation location within a given world view. This does not, of course, mean that it does not require solution.

24. Toward a Strategy for Solution of the Free-Will Problem in Ethics

A strategy for solution of the problem of free will and responsibility in the light of the kind of study mapped above is primarily one of separating the strands and relating them either to the area of ethical theory to which they are found connected or else to the specific EP from which they stem and the fate of whose evaluation they share. Many strands furnish what turn out to be answers to some part of the whole network of problems.

(i) In some pragmatic accounts, to find a man responsible is not to attribute to him some antecedent voluntaristic phenomenon but to affirm that he ought to be punished or rewarded. The reasons may be various, dependent on field, purposes of the enterprise, general values.[116] Whether this is an adequate general account or not, it does call attention to one strand: that some large part of responsibility theory can be seen as a vehicle of standard-formation for the application of sanctions. It even verges in some cases—for example, in assigning liability for accidents which are "nobody's fault"—on a simple working-out of principles of distribution of gains and losses as one works out principles of taxation.

(ii) We can look in traditional accounts of freedom for the portrait of the free man—whether the Stoic with complete self-possession, the Spinozistic free man in a determinist world with clear knowledge of his aims and conditions and viewing his actions in terms of determination from the essential nature of the whole, the Kantian conscientious universalistic self-regulator, the Marxian free man as fully conscious of necessity as possible in his time and turning his knowledge into social action for the further achievement of human values, the Russellian free man defying matter and worshiping at the shrine of his own values, the Freudian free man with insight into himself strengthening his ego to extend its sphere, or a host of others.[117] We can separate them from the issues of belief in determinism or metaphysical free will and see how far they function as proposals for virtue configurations, to be assessed as ethical standards.

(iii) Some judgments of coercion, intimidation, provocation, etc., can also be seen as ways of determining permissible processes in institutional and interpersonal relations. Deciding what is or is not coercion in contractual relations, or in labor relations, or in the relation of superior and inferior in the military organization may in effect be valuational decision on how far the state may interfere to provide social security, what permissible weapons there may be in business-labor tension, what limits there are to obedience to commands of superiors (cf. the problems of responsibility in the Nuremberg trials).

(iv) Other judgments of coercion or compulsion center on the voluntaristic phenomena and attempt to establish or invoke empirical criteria of assent or extent of internal harmony in decision. Schlick went so far as to assert that the issue of free will versus determinism was a pseudo-issue and that voluntary decision versus compulsion from others was the only issue that made sense.[118] In any case, it is one of the issues, and its scope is broadened by psychological knowledge of compulsions, of unconscious influence of authority-figures, etc. Similarly, increased social knowledge of the strength of social forces can help stabilize the indices of coercion.

(v) Part of the traditional free-will controversy concerned primarily whether man is to be studied as part of nature or is in some sense outside the order of nature. Contemporary scientific philosophies regard this issue as settled by this time, and the full incorporation of man within the natural world as established.

(vi) Part of the argument for free will has centered on the possibility of novelty in the world, which seemed to be ruled out by a complete or hard determinism. In one form or another, the advance of the philosophy of science has furnished analyses in which complete predictability to the last fine shade is not an essential mark of determinism and in which the inability to predict in *advance* does not entail the inability to find causes *after* the novel phenomenon has come into existence.[119] (In the theory of mind, this has meant that a scientific approach is not inherently bound to epiphenomenalism.)

(vii) A step beyond issues of novelty have come attempts to work out a principle of creativity as somehow anti-deterministic. Here, again, once the possibility of novelty is granted, the phenomena of creativity become detached from specific free-willist doctrine. In empirical terms it is possible to map the points of creativity in biological evolution, in individual biography, in history. That this is not incapable of association with a causal approach can be seen in Marxian materialism, in which freedom is identified as the growth of effective consciousness in human life with the evolution of the material-social world.

(viii) Some stress the phenomenological datum of the sense of freedom in the first-person act of decision, as wholly removed from scientific scrutiny and explanation. This can, however, be seen as a problem in the psychological theory of the self. It is by no means solved at the present stage of inquiry and may itself prove to contain a fusion of strands or cover a multiple rather than a single issue. Certainly one would have to distinguish the feeling of the reality of the act of will (the sense that a self or an I is involved in the process), the consciousness of the I as the source or starting-point of the act, the sense of efficacy of the will (that something different or novel may come from the fact that I willed this rather than that), the sense of possible alternatives (that I could have willed otherwise), and perhaps many other elements. In spite of the numerous attempts to extract a transcendental ego from these phenomenological data, there appears to be nothing here that a growing conception of the self as a special kind of dynamical system in the career of the individual organism may not take in its stride. Naturalistic theories have tried to exhibit the sense of freedom sometimes as a sign of the congruity of the path chosen with the basic desires of the agent, sometimes as the sense of absence of coercion by others, sometimes as the cognitive recognition that the act was in some quite specific sense "avoidable." Another possibility, as yet insufficiently explored, is that the situation of choice is itself structured in such a way as (phenomenologically) to create a gap (cognitive "distance") between the self and the field, so that determinism cannot be taken to apply to the self without putting the self into the field.[120] Hence, *in the act of choosing,* the sense of freedom would be phenomenologically meaningful; but in the subsequent study of his choice even by the agent himself, determinism would be likewise meaningful. At the present stage of the growth of the psychological study of the self, what is needed is a theory that will serve to relate the organic, the behavioral, the phenomenological, the developmental, and even the cultural and the social approaches.

How far the strategy here envisaged may serve to solve the complex problems in the freedom and responsibility area depends of course on the results of the researches envisaged. The very least we may expect are a broader presentation of the phenomena and a breakthrough in the stalemate of the traditional issue.

25. The Creative Temper and the Scientific Temper

At the opening of this work we posed the problems of the place of scientific results, the role of scientific method, and the impact of the scientific temper in ethical theory. Scientific results have been found to have an unavoidable place in the understanding of ethical theories and an indispensable position in their evaluation. The utilization of scientific method in ethics has been shown to be genuinely possible, with the extent of its practicability dependent on the extent of determinateness found in the human field. It has not been the task of this work to carry out an inventory of the scientific materials that establish the requisite extent of this determinateness. The policy recommendation that scientific method be utilized in ethics—as distinct from the logical and methodological anlysis of its possibility—involves a recognition of the understanding the growth of the sciences of man has already brought to ethics. There is also the promise of further understanding with the further growth of these inquiries, especially when directed to the phenomena of the moral field itself. And there is the recognition that alternative policies have not proved fruitful in the development of ethical theory and the suspicion that too often they themselves rest on some *ersatz* science.

Decision about the scientific temper is a further question. In part, of course, it would be necessitated in a policy decision to utilize scientific method in ethics. But that this advance-guard of the scientific outlook should be allowed to permeate the field of ethical inquiry, no matter how far and in what detail scientific method be utilizable, is itself an ethical question. For the scientific temper would function in ethics as a set of character-traits and that is as a virtue-constellation. To assess its impact is therefore to estimate both what has been and what ought to be the place of these virtues in ethics. It is, in short, part of the theory of virtues in a morality.

Since we have not attempted to go into the field of evaluative morality in this work, the full answer lies beyond our present scope. There is, however, a basic consideration which may be offered by way of recommendation for these virtues. Man's whole development is tied up with the progress of knowledge and its application; the virtues of reflective assessment of experience, wide range of investigation, sober weighing of evidence, and the rest of the familiar list have certainly earned their keep. What is more, ethical theory itself has always been a reflective enterprise, with rare and impassioned exceptions. It might be well to try out the explicit development of the whole field in a scientific spirit.

Perhaps the chief criticisms of the scientific temper have centered on misunderstandings of its nature—the contrast between the scientific and the creative. This reflects what is sometimes the parochial scientism of a limited period, the blind emphasis on the mechanical and the measurable. A less stereotyped view of the scientific temper is to be found in its greatest geniuses and in the history of science at critical periods. There it is the imaginative, the grasp of new possibilities, the ability to see things never seen before and to pose questions in a new way, to strip off blinders rather than to impose a new brand of them, which strike the investigator. Then the scientific temper and the creative temper become two faces of a single coin. It would indeed be a strange retribution if mankind, so prone to seek its salvation in the act, to conjure up romanticisms of the heart and the will, were to find the stoutest ally for both heart and will in the quest for knowledge.

Notes

CHAPTER I

1. I should like to express my indebtedness to the National Science Foundation for a grant, during the academic year 1959–60, to continue my work on the relation of science and ethics. The present monograph is a part of this work.

2. Cf. Abraham Edel, *Ethical Judgment: The Use of Science in Ethics* (Glencoe, Ill.: Free Press, 1955), chaps. v–viii.

3. Exceptions will be found in Charles Morris, *Signs, Language and Behavior* (New York: Prentice-Hall, Inc., 1946), pp. 230–33; DeWitt H. Parker, *The Philosophy of Value* (Ann Arbor: University of Michigan Press, 1957), p. 87. For general discussion of this problem cf. Abraham Edel, *Method in Ethical Theory* (New York: Liberal Arts Press, 1961), chap. viii.

4. Cf. May Edel and Abraham Edel, *Anthropology and Ethics* (Springfield, Ill.: Charles C Thomas, 1959).

5. My *Method in Ethical Theory* is devoted to an exposition of these four standpoints.

6. Cf. Erich Fromm, *Man for Himself* (New York: Rinehart & Co., 1947), pp. 141–72.

CHAPTER II

7. Cf. Edel and Edel, *op. cit.*, chap. xiv.

8. Reference to some types of existentialist perspectives will be found in section 14 below.

9. Clyde Kluckhohn and others, "Values and Value-Orientation in the Theory of Action," in Talcott Parsons and Edward A. Shils (eds.), *Toward a General Theory of Action* (Cambridge, Mass.: Harvard University Press, 1951), p. 411.

10. Robert Redfield, *The Primitive World and Its Transformations* (Ithaca, N.Y.: Cornell University Press, 1953), pp. 85, 86. See the whole of chap. iv.

11. Cf. Daryll Forde (ed.), *African Worlds: Studies in the Cosmological Ideas and Social Values of African Peoples* (New York: Oxford University Press [for the International African Institute], 1954); Ethel Albert, "The Classification of Values: A Method and Illustration," *American Anthropologist*, LVIII (April, 1956), 221–48; Florence Kluckhohn, "Dominant and Substitute Profiles of Cultural Orientations: Their Significance for the Analysis of Social Stratification," *Social Forces*, XXVIII (1950), 376–93; A. I. Hallowell, *Culture and Experience* (Philadelphia: University of Pennsylvania Press, 1955), esp. chap. iv, and chaps. viii, ix, and xi; John J. Honigmann, *The World of Man* (New York: Harper & Bros., 1959), chaps. xxxiv–xxxix.

12. The stage-setting concept was worked out for ethics in my "Coordinates of Criticism in Ethical Theory," *Philosophy and Phenomenological Research*, VII (June, 1947), esp. pp. 550–54 (reprinted in Edel, *Method in Ethical Theory*). It was suggested by A. F. Bentley's procedure in schematizing psychological approaches in his *Behavior, Knowledge, Fact* (Bloomington. Ind.: Principia Press, 1935), esp. Part I. Cf. also the mode of schematic representation in Egon Brunswik, *The Conceptual Framework of Psychology*, in this *Encyclopedia*, Vol. I, No. 10.

13. Cf. Meyer Fortes, *Oedipus and Job in West African Religion* (Cambridge: Cambridge University Press, 1959).

14. Jeremy Bentham, *Principles of Morals and Legislation* (new edition corrected by author, 1823; republished, Oxford: Clarendon Press, 1907); J. S. Mill, *Utilitarianism* (1863) (reprinted, New York: Liberal Arts Press, 1949).

15. Thomas Hobbes, *Leviathan* (London, 1651).

16. Cf. Herbert Spencer, *Social Statics* (1850) (New York: D. Appleton & Co., 1865); Peter Kropotkin, *Mutual Aid* (1902) (Penguin Books, Ltd., 1939).

17. Cf. Leslie Stephen, *The Science of Ethics* (New York: G. P. Putnam's Sons, 1882), esp. chap. iii.

18. Herbert Spencer, *The Principles of Ethics* (New York: D. Appleton & Co., 1896–97), Vols. I–II.

19. Adam Smith, *The Theory of the Moral Sentiments* (London, 1759; 6th ed., 1790), Part III (reprinted in part in *Adam Smith's Moral and Political Philosophy* [New York: Hafner Publishing Co., 1948]).

20. Bentham, *op. cit.*, chap. iv.

21. Bertrand Russell, *Human Society in Ethics and Politics* (New York: Simon & Schuster, 1955).

22. Moritz Schlick, *Problems of Ethics* (New York: Prentice-Hall, Inc., 1939), chap. ii.

23. Ralph Barton Perry, *General Theory of Value* (New York: Longmans, Green & Co., 1926), p. 209.

24. *Ibid.*, p. 471.

25. E. C. Tolman, *Purposive Behavior in Animals and Men* (New York: Century Co., 1932); see also the collected papers in *Behavior and Psychological Man* (Berkeley: University of California Press, 1958).

26. Stephen C. Pepper, *The Sources of Value* (Berkeley: University of California Press, 1958).

27. For a general picture see Sigmund Freud, *New Introductory Lectures on Psycho-analysis* (New York: W. W. Norton & Co., 1933). The biological, social, and historical theses are found, respectively, in *Beyond the Pleasure Principle* (New York: Boni & Liveright), *Group Psychology and the Analysis of the Ego* (New York: Boni & Liveright), and *Civilization and Its Discontents* (New York: Jonathan Cape & Harrison Smith, 1930).

28. E.g., Heinz Hartmann, "Ego Psychology and the Problem of Adaptation," in *Organization and Pathology of Thought*, ed. David Rapaport (New York: Columbia University Press, 1951).

29. Such theories are found in Freudian, neo-Freudian, and general psychological schools (cf. Otto Fenichel, *The Psychoanalytic Theory of Neurosis* [New York: W. W. Norton & Co., 1945], pp. 59–61; Fromm, *op. cit.*, pp. 210–26; and John Dollard *et al.*, *Frustration and Aggression* [New Haven, Conn.: Yale University Press, 1939]).

30. Cf. Fromm, *op. cit.*; Karen Horney, *The Neurotic Personality of Our Time* (New York: W. W. Norton & Co., 1937); Erik H. Erikson, *Childhood and Society* (New York: W. W. Norton & Co., 1950); J. C. Flugel, *Man, Morals and Society* (New York: International Universities Press, 1945); Abram Kardiner, *The Psychological Frontiers of Society* (New York: Columbia University Press, 1945).

31. Harry Stack Sullivan, *Conceptions of Modern Psychiatry* (2d ed.; New York: W. W. Norton & Co., 1953).

32. Among psychologists cf. Wolfgang Köhler, *The Place of Value in a World of Facts* (New York: Liveright Publishing Corp., 1938), chap. iii; Karl Duncker, "Ethical Relativity? (An Enquiry into the Psychology of Ethics)," *Mind*, XLVIII (January, 1939), 39–57; Solomon E. Asch, *Social Psychology* (New York: Prentice-Hall, Inc., 1952), chaps. ii, xii, and xiii. Among philosophers, Maurice

Mandelbaum, *The Phenomenology of Moral Experience* (Glencoe, Ill.: Free Press, 1955); Nicolai Hartmann, *Ethics* (New York: Macmillan Co., 1932), Vols. I–III; and Max Scheler, *The Nature of Sympathy* (London: Routledge & Kegan Paul, Ltd., 1954). Hartmann and Scheler go beyond descriptive phenomenology, with which alone we are here concerned.

33. Cf. sections on phenomenological description in Edel, *Method in Ethical Theory*, chaps. viii and xi.

34. Cf. Charles L. Stevenson, *Ethics and Language* (New Haven, Conn.: Yale University Press, 1944), chaps. i, ii.

35. John Dewey, *Human Nature and Conduct* (New York: Modern Library, 1930), p. 39.

36. In an article, "Some Questions about Value?" *Journal of Philosophy*, LI (August 17, 1944), 455, Dewey asks: "Are values and valuations such that they can be treated on a psychological basis of an allegedly 'individual' kind? Or are they so definitely and completely socio-cultural that they can be effectively dealt with only in that context?"

37. John Dewey, in Part II of Dewey and Tufts, *Ethics* (rev. ed.; New York: Henry Holt & Co., 1932). Cf. his concluding statement (pp. 342–44).

38. The outlines of the Marxian theory of ethics are clear in Frederick Engels, *Anti-Duehring* (New York: International Publishers), chaps. ix–xi and Part III.

39. Spencer's mode of analysis is clear in the detailed discussion of his *The Principles of Ethics*. For Kropotkin see *op. cit.*, chaps. iii–viii; also his *Ethics, Origins and Development* (New York: Dial Press, 1924).

40. Julian Huxley, "Evolutionary Ethics," in T. H. Huxley and Julian Huxley, *Touchstone for Ethics* (New York: Harper & Bros., 1947).

41. *Immanuel Kant's Critique of Pure Reason*, trans. Norman Kemp Smith (London: Macmillan & Co., 1929), p. 643. For illustration of Kant's detailed elaboration of his approach see, e.g., *Immanuel Kant's Religion within the Limits of Reason Alone*, trans. T. M. Greene and H. H. Hudson (Chicago: Open Court Publishing Co., 1934), p. 131.

42. The kinds of arguments used in traditional doctrinal conflicts constitute illuminating materials for such an inquiry (cf. *Documents of the Christian Church*, ed. Henry Bettenson [New York: Oxford University Press, 1947]).

43. *International Journal of Ethics*, October, 1903, p. 116.

44. G. E. Moore, *Principia ethica* (Cambridge: Cambridge University Press, 1903), chap. i.

45. Compare John Laird's comment in his *An Enquiry into Moral Notions* (London: George Allen & Unwin Ltd., 1935), p. 106: "What Kant called 'respect' for the moral law is a ghost from Sinai, a crepuscular thing that sins against the natural light. Therefore consistent Kantians must either bring divinity into their ethics, not as a consequence but as part of the analytic of their fundamental conceptions, or else retire to purely terrene ramparts and abandon their view that the moral fact is unique of its kind."

46. T. H. Green, *Prolegomena to Ethics* (Oxford: Clarendon Press, 1883). The issue is clearly posed in the Introduction.

47. Other clues occasionally useful in detecting an implicit EP are: conceptions of the tasks of ethics, contexts of application, guiding models.

48. Moore, *op. cit.*, pp. 143 ff.; Hartmann, *op. cit.*, I, 218 ff.

49. Reinhold Niebuhr, *The Nature and Destiny of Man* (New York: Charles Scribner's Sons, 1943), I, 13–14.

50. Jean-Paul Sartre, *The Transcendence of the Ego* (New York: Noonday Press, 1957); *Existentialism and Humanism* (London: Methuen & Co., 1948).

51. Niebuhr, *op. cit.*, I, 122–25; George Santayana, *Reason in Society* (New York: Charles Scribner's Sons, 1930), pp. 3–4; Gilbert Ryle, *The Concept of Mind* (London: Hutchinson's University Library, 1949), pp. 195–98.

52. For a detailed criticism along these lines see Abraham Edel, "The Logical Structure of G. E. Moore's Ethical Theory," in *The Philosophy of G. E. Moore*, ed. P. A. Schilpp ("Library of Living Philosophers," Vol. IV [Evanston: Northwestern University, 1942]), esp. pp. 169–76.

53. Karen Horney, *New Ways in Psychoanalysis* (New York: W. W. Norton & Co., 1939), pp. 187–88.

54. William James, *Pragmatism* (1907) (New York: Longmans, Green & Co., 1947), p. 119.

55. For a general discussion of different relations in which ideas can stand to social context see Abraham Edel, "Context and Content in the Theory of Ideas," in *Philosophy for the Future*, ed. R. W. Sellars, V. J. McGill, and M. Farber (New York: Macmillan Co., 1949).

56. For an attempt to work out sociocultural categories for structuring ethical theory see Edel, *Method in Ethical Theory*, chap. x.

CHAPTER III

57. Cf. Edel, "Coordinates of Criticism in Ethical Theory," *op. cit.*

58. Cf. R. B. Brandt, "The Status of Empirical Assertion Theories in Ethics," *Mind*, LXI (October, 1952), 458–79.

59. Cf. Edel, *Method in Ethical Theory*, chap. vii, sec. 6.

60. Fyodor Dostoevski, *Letters from the Underworld* ("Everyman" ed.; New York: E. P. Dutton & Co., 1945), p. 27.

61. For an examination of the problem of the mark of the moral, and for consideration of different candidates, see Edel and Edel, *op. cit.*, chap. ii; also *Method in Ethical Theory*, chap. viii, section on "The Data of Ethics."

62. For a study of the three families and some problems in their relation see Laird, *op. cit.*

63. For a study of the variety that may be embraced in the notion of defining ethical terms cf. Edel, *Method in Ethical Theory*, chap. vi.

64. For Socratic and Platonic views, Plato's *Apology, Crito, Symposium*. For Stoics, Epictetus' *Enchiridion* and Marcus Aurelius' *Meditations*. For Kant, his *Critique of Practical Reason* and his *Fundamental Principles of the Metaphysic of Morals;* his *Lectures on Ethics* (London: Methuen & Co., 1930) (compiled out of students' lecture-notes) contains illuminating materials on the psychology of morals. For Nietzsche, his *Genealogy of Morals*. For Bentham, Köhler, Mandelbaum, works cited in previous notes. For Freud, works cited in n. 27, above. Also Bernard Bosanquet, *The Principle of Individuality and Value* (London: Macmillan & Co., 1912), Lecture IV; Josiah Royce, *The Philosophy of Loyalty* (New York: Macmillan Co., 1911); Henri Bergson, *The Two Sources of Morality and Religion* (New York: Henry Holt & Co., 1935), chap. i.

65. Lewis Samuel Feuer, *Psychoanalysis and Ethics* (Springfield, Ill.: Charles C Thomas, 1955), p. 21.

66. Ruth Benedict, *The Chrysanthemum and the Sword* (New York: Houghton Mifflin Co., 1946), chaps. v–vii; see especially the schematic table of obligations, p. 116.

67. *Ibid.*, p. 102.

68. P. H. Nowell-Smith, *Ethics* (London: Penguin Books, 1954), chap. vii.

69. See works, cited in previous notes, by Bentham, Bosanquet, Dewey, Parker, Pepper, Perry. For homeostatic models cf. Anatol Rapoport, "Homeostasis Re-

considered," in Roy R. Grinker (ed.), *Towards a Unified Theory of Human Behavior* (New York: Basic Books, 1956).

70. It is sometimes held, however, that an authoritative particular command requires justification by reference to universals. This would seem to depend on the meaning of 'authoritative'; also, it may be that the demand for a universal element creeps in through the notion of justification. For a defense of the universalist condition see R. B. Brandt, *Ethical Theory* (Englewood Cliffs, N.J.: Prentice-Hall, Inc., 1959), chap. ii.

71. These are presented in greater detail in Edel, *Ethical Judgment*, chap. ii.

72. The concept of *phase rule* is intended to meet the kind of problems for which W. D. Ross employs the notion of *prima facie duty* (*The Right and the Good* [Oxford: Clarendon Press, 1930], pp. 24–29); or John Dewey distinguishes *principle* from *rule* (Dewey and Tufts, *op. cit.*, pp. 304–5); or, in terms of differentiating effect within the individual's consciousness, Hartmann develops the notion of *unavoidable guilt* (*op. cit.*, I, 299–302; cf. II, 281–85).

73. Ralph Linton, *The Study of Man* (New York: D. Appleton–Century Co., 1936), p. 433. The context is a discussion of patterns of misconduct.

74. Cf. P. F. Strawson, "Construction and Analysis," in A. J. Ayer *et al.*, *The Revolution in Philosophy* (London: Macmillan & Co., 1956).

75. Cf. Edel, *Method in Ethical Theory*, chap. v.

76. For Kant see works cited above, n. 64; Schlick, *op. cit.*, p. 32; Edward Westermarck, *Ethical Relativity* (New York: Harcourt Brace & Co., 1932), pp. 152–53; Ross, *op. cit.*, pp. 6–7. The view that there is no restriction on the subject in the case of 'good' appears to be G. E. Moore's in *Principia ethica*, although later (in "Is Goodness a Quality?" *Aristotelian Society*, Supplementary Volume XI [1932], 122–24) he suggests translating 'good' into 'worth having for its own sake,' which limits the subject to experiences. The hedonist would, of course, share this limitation for different reasons.

77. Ross, *op. cit.*, pp. 6–7; *Foundations of Ethics* (Oxford: Clarendon Press, 1939), p. 115.

78. In modern ethics the traditional contrast of Utilitarian and Kantian ethics illustrates the contrast of goodist and rightist frameworks. In twentieth-century ethics G. E. Moore's *Principia ethica* is a clear illustration of the goodist type; see also H. W. B. Joseph, *Some Problems of Ethics* (Oxford: Clarendon Press, 1931). The insistence on primacy for 'right' or 'ought' is found in H. A. Prichard, *Moral Obligation* (Oxford: Clarendon Press, 1949), especially Essay 5. Hartmann's translation of value into an Ought-to-Be carries a rightist connotation (*op. cit.*). Recent works sometimes make a formal translation along these lines; e.g., Everett W. Hall, in *What Is Value?* (London: Routledge & Kegan Paul Ltd., 1952) considers '*a* is good' as meaning 'There is a property, X, such that *a* ought to exemplify X and *a* does exemplify X' (p. 178). The coordinate structure is a basic thesis in Ross (works cited in nn. 72 and 77). Insistence on fundamental separation of value and obligation is found in C. I. Lewis, *An Analysis of Knowledge and Valuation* (LaSalle, Ill.: Open Court Publishing Co., 1946), chaps. xiii and xvi. It takes the form of a separation of evaluation and obligation as two functions of language, in Alexander Sesonske's *Value and Obligation* (Berkeley: University of California Press, 1957). A different mode of separation (between theory of duties and theory of ideals) is found in Leonard Nelson's *System of Ethics* (New Haven, Conn.: Yale University Press, 1956). Dewey makes an interesting attempt at reconciling the conflict of the right and the good (Dewey and Tufts, *op. cit.*, pp. 249 ff.).

79. Nowell-Smith, *op. cit.*, p. 133.

80. Westermarck, *op. cit.*, p. 122.

81. Cf., respectively, Rudolf Carnap, *Philosophy and Logical Syntax* (London: Kegan Paul, 1935), pp. 22–26; A. J. Ayer, *Language, Truth and Logic* (London: Victor Gollancz, 1936), chap. vi; Bertrand Russell, *Religion and Science* (New York: Henry Holt & Co., 1935), p. 235.

82. E.g., Hall, *op. cit.*, esp. chap. 6; R. M. Hare, *The Language of Morals* (Oxford: Clarendon Press, 1952), chap. xii. See also n. 83.

83. For a useful brief outline of this area see A. R. Anderson and O. K. Moore, "The Formal Analysis of Normative Concepts," *American Sociological Review,* XXII (February, 1957), 9–17. For a system embodying 'better' as fundamental concept see Soren Halldén, *On the Logic of 'Better'* ("Library of Theoria," No. 2 [Lund: C. W. K. Gleerup, 1957]).

84. E.g., Anderson and Moore say: "Certain writers have evinced an extreme reluctance, for philosophical and grammatical reasons, to admit any very close relation between propositions and commands. In our opinion this is an open question, to be decided by constructing logical systems whose utility can be tested in scientific practice" (*op. cit.*, p. 15). They take as their point of departure the suggestion by H. G. Bohnert ("The Semiotic Status of Commands," *Philosophy of Science,* XII [July, 1945], 302–15) that "Do A" is to be taken as elliptical for "Either you will do A, or else S," where S refers to a sanction. For a criticism of Bohnert's view see Hall, *op. cit.*, pp. 131–32, and Hare, *op. cit.*, pp. 7–8. There is an interesting parallel between this kind of logical attempt and Justice Holmes's attempt to construe a contract as assumption of risk, that is, a prediction that something will come to pass or else certain "damages" will be assessed (see Oliver Wendell Holmes, Jr., *The Common Law* [1881] [Boston: Little, Brown & Co., 1938], pp. 298 ff.).

85. Edel and Edel, *op. cit.*, pp. 127–29 and chap. xiv.

86. Kurt Baier, *The Moral Point of View* (Ithaca, N.Y.: Cornell University Press, 1958), p. 183.

87. Moore, *Principia ethica,* pp. 97–99.

88. Henry Sidgwick, *The Methods of Ethics* (6th ed.; London: Macmillan & Co. Ltd., 1901), pp. 381 ff. Cf. C. I. Lewis, *op. cit.*, pp. 483 ff., 492 ff.

89. Moore, *Principia ethica,* pp. 143–44, quite clearly differentiates his view from much of traditional intuitionism.

90. Duncker, *op. cit.*

91. Stevenson, *op. cit.*, chap. vii.

92. This term is Gilbert Ryle's. For his conception of informal logic see his *Dilemmas* (Cambridge: Cambridge University Press, 1956), chap. viii.

93. Stephen E. Toulmin, *The Place of Reason in Ethics* (Cambridge: Cambridge University Press, 1953).

94. Baier, *op. cit.*, pp. 86–87.

95. *Ibid.*, pp. 142–43.

96. *Ibid.*, p. 260.

97. *Ibid.*, p. 301.

98. A parallel analysis would hold if we give an activist interpretation to furnishing good reasons. In that case the linguistic formulation would be serving the purpose of "announcing a policy of action" (Stuart Hampshire, *Thought and Action* [London: Chatto & Windus, 1959], p. 129). Hampshire adds: "I do not helplessly encounter reasons for action; I acknowledge certain things as reasons for action." The relation between phenomenological and activist components is a separate question.

99. Cf. J. O. Urmson's attempt to interpret 'good' as a grading label, in his "On Grading," *Mind,* LIX (1950), 145–69.

100. Bentham's felicific calculus is a classic source in the search for dimensions

of value measurement (*op. cit.*, chap. iv). For attention to these problems in value theory cf. Perry, *op. cit.*, chap. xxi, and his *Realms of Value* (Cambridge, Mass.: Harvard University Press, 1954), chap. iv; John Laird, *The Idea of Value* (Cambridge: Cambridge University Press, 1929), chap. x; Parker, *op. cit.*, chaps. v–vii; Charles Morris, *Varieties of Human Value* (Chicago: University of Chicago Press, 1956). While generic value criteria have been explored, criteria in the specific moral field tend to be overlooked, often on the assumption that they are special applications of the generic.

101. Edel, *Method in Ethical Theory*, chap. xvi.

102. Erikson, *op. cit.*, p. 218 and chap. vii; cf. also pp. 230–34.

103. Parker, *op. cit.*, pp. 152–53. He models this on the touchstone method of criticism in literature, expounded by Matthew Arnold.

104. R. B. Braithwaite, *Moral Principles and Inductive Policies* (Annual Philosophical Lecture, Henriette Herz Trust, British Academy, 1950 [*Proceedings of the British Academy* (London), XXXVI, 68]).

105. Compare the way in which C. I. Lewis casts his basic axiom for obligation (*op. cit.*, pp. 503–10).

106. Edel, *Ethical Judgment*, chap. ix.

CHAPTER IV

107. Cf. the works of Dewey cited in nn. 35 and 37, above. See also his *Theory of Valuation*, in this *Encyclopedia*, Vol. II, No. 4.

108. Sartre, *Existentialism and Humanism*.

109. For a recent presentation of views cf. *Determinism and Freedom in the Age of Modern Science: A Philosophical Symposium*, ed. Sidney Hook (New York University Press, 1958); cf. also Wilfrid Sellars and John Hospers (eds.), *Readings in Ethical Theory* (New York: Appleton-Century-Crofts, Inc., 1952), Part VII; Herbert Fingarette, "Psychoanalytic Perspectives on Moral Guilt and Responsibility: A Re-evaluation," *Philosophy and Phenomenological Research*, XVI (September 1955), 18–36.

110. Flugel (*op. cit.*) discusses such materials with special reference to morals.

111. Smith, *The Theory of the Moral Sentiments*, p. 135.

112. For a brief sketch see Roscoe Pound, *An Introduction to the Philosophy of Law* (New Haven, Conn.: Yale University Press, 1922), chap. iv.

113. Hartmann, *op. cit.*, I, 299–302.

114. For an interesting treatment of the growth of this individualism in relation to the legal category of contract cf. Morris R. Cohen, *Law and the Social Order* (New York: Harcourt, Brace & Co., 1933), pp. 74–88.

115. For some of the complexities in the analysis of shame see Gerhart Piers and Milton B. Singer, *Shame and Guilt* (Springfield, Ill.: Charles C Thomas, 1953). For the impact of shame in the growth of an individual see Helen M. Lynd, *On Shame and the Search for Identity* (New York: Harcourt, Brace & Co., 1958). For blame see Herbert Fingarette, "Blame: Its Motive and Meaning in Everyday Life," *Psychoanalytic Review*, XLIV (April, 1957), 193–211.

116. J. S. Mill's *Utilitarianism*, chap. v, inclines at several points to such a mode of analysis. For an explicit application to the theory of praise and blame see Dewey, *Human Nature and Conduct*, e.g., pp. 314 ff. For a parallel mode of thought in legal theory cf. Oliver Wendell Holmes, "The Path of the Law," in *Collected Legal Papers* (New York: Harcourt, Brace & Co., 1920).

117. Benedict de Spinoza, *Ethics*, Part V; Bertrand Russell, "A Free Man's Worship," in *Mysticism and Logic* (New York: W. W. Norton & Co., 1929). For Stoic view see n. 64 above; Kantian, n. 64; Marxian, n. 38; Freudian, n. 27.

118. Schlick, *op. cit.*, chap. vii.

119. For consideration of the problems of novelty and emergence cf. A. O. Lovejoy, "The Meanings of 'Emergence' and Its Modes," *Proceedings of the Sixth International Congress of Philosophy* (1926), pp. 20–33; Ernest Nagel, "The Meaning of Reduction in the Natural Sciences," in Robert C. Stauffer (ed.), *Science and Civilization* (Madison: University of Wisconsin Press, 1949); Abraham Edel, *The Theory and Practice of Philosophy* (New York: Harcourt, Brace & Co., 1946), pp. 48–64; Gustav Bergmann, "Holism, Historicism and Emergence," *Philosophy of Science*, XI (October, 1944), 209–21; P. E. Meehl and Wilfrid Sellars, "The Concept of Emergence," in H. Feigl and M. Scriven (eds.), *The Foundations of Science and the Concepts of Psychology and Psychoanalysis* (Minneapolis: University of Minnesota Press, 1956), pp. 239–52; Paul Oppenheim and Hilary Putnam, "Unity of Science as a Working Hypothesis," in H. Feigl, M. Scriven, and G. Maxwell (eds.), *Concepts, Theories and the Mind-Body Problem* (Minneapolis: University of Minnesota Press, 1958), pp. 3–36; R. W. Sellars, V. J. McGill, and M. Farber (eds.), *Philosophy for the Future* (New York: Macmillan Co., 1949), essays by J. B. S. Haldane, T. C. Schneirla, and B. J. Stern.

120. Gilbert Ryle, in the reference cited above, n. 51, seems to me to be developing such a conception, though in different terms.

Milton Keynes UK
Ingram Content Group UK Ltd.
UKHW022109141024
449569UK00031B/1845